CW01668655

Calculator Calculus

Barrie Galpin and Alan Graham

A+B Books

Written for the Texas Instruments TI-83 family of graphics calculators.

First published 2002 and reprinted 2003.

ISBN 0-9541020-2-9

Published by A+B Books, 15 Top Lodge, Fineshade, Near Corby NN17 3BB.
www.AplusB.co.uk

Cover design: Jeanette Ward.
Cartoons: original work by Steve Smith.
Printed in the UK by the Direct Printing Company Ltd, Brixworth, Northants.

Getting started

This book forms part of a series aimed at students of 16 years and over. It is written for use with the Texas Instruments TI-83 family of graphics calculators and makes use of their ability to carry out numerical differentiation and integration. Such calculators do not carry out algebraic manipulation – they cannot differentiate and integrate functions directly. However, as you will see, they can be used to develop a firm understanding of the underlying ideas of calculus.

If you are meeting calculus for the first time or coming back to it after years of lack of use, then this book, **Calculator Calculus** is the one for you.

However, you do need to have a certain level of skill with a graphics calculator before you begin. If you are using the calculator for the first time or are not too confident about using it, then consider working through **Calculator Maths: Foundations Plus** first.

Calculator Calculus tackles the two big ideas of Calculus: differentiation and integration. The first ten units of this book deal with differentiation, while the final four cover the related topic of integration. The units provide a fresh way of viewing these topics – ideal for using alongside traditional text books or as a means of revising previously covered material.

You will find that it is best to study the units in the order in which they appear – generally each unit builds on those that precede it.

Each unit consists of a number of *activities*. Think about that word – you are meant to be active as you do them and that means an active head as well as active fingers! As you do each activity you will be learning and gaining in mathematical understanding.

At the back of the book are quite detailed comments and solutions. You are sure to want to use the answers to check your work, especially if you are a student working alone without a teacher.

Most units also have some *Brain stretchers*. They are optional and usually harder sections that may stretch your brain just that bit further. Does your brain need stretching? – if so, try some of these.

Calculator Maths: Foundations Plus has been written to provide new users with all the basic calculator skills they are likely to need. The book provides:

- an introduction to the calculator with practice exercises,
- a useful reference sectionwhere you can look up particular calculator skills and methods if you need to,
- a section on *Writing your own programs*,
- help with *Trouble Shooting* if things do go wrong.

Calculator Maths: Foundations Plus is also written by Alan Graham and Barrie Galpin and is published by and available from A+B Books.

Contents

Unit 1 Gradient of a straight line

Underlying the big mathematical idea of differentiation is the concept of a gradient, that is the means of measuring steepness.

The aim of this unit is to reinforce your understanding of the meaning of gradients in mathematics and introduce calculator techniques, such as drawing graphs and creating programs, which you will need later in this book.

The word *gradient* is one of many used in everyday speech that have a very special meaning in mathematics – other examples are *similar, product* and *differentiate.* As you work through the unit, look out for ways in which the term gradient is being used that are different from those used in everyday life.

(1) The starting line

Let's start by drawing a straight line on your calculator's screen. Make sure that you have no statistical plots selected or functions entered on the Y= screen – see page 16 of **Calculator Maths: Foundations Plus** if you are not sure how to do this.

(a) Choose some suitable axes – there are several pre-programmed ones in the ZOOM menu and **ZDecimal** is suitable here. So press [ZOOM] **4**.

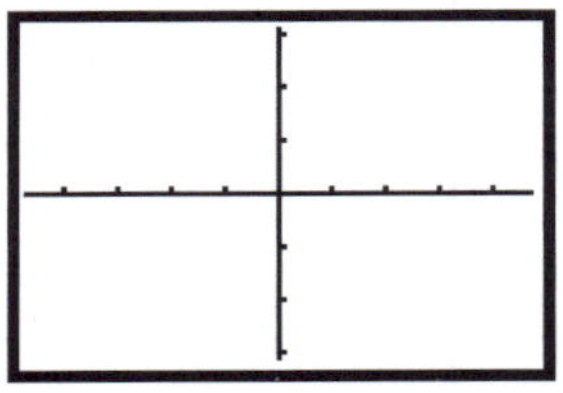

You can see a pair of axes on the graphing screen. The x-axis runs from about $^-5$ to 5 and the *y*-axis from about $^-3$ to 3.

(b) To draw on the graphing screen you use commands from the DRAW menu. However, these commands operate differently depending on whether the graphing screen or home screen is currently selected.

So, starting from the graphing screen, choose the option that lets you draw lines by pressing

[2nd] [DRAW] **2**.

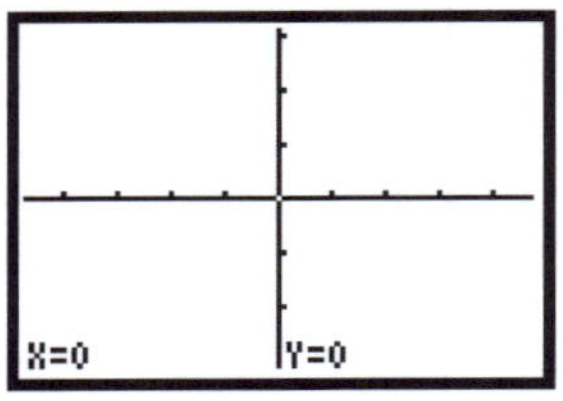

Look carefully at the subtle differences on the graphing screen from before. A small flashing cursor has appeared at the origin (where the two axes cross) and its coordinates are shown below: X=0 and Y=0.

(c) Press [ENTER] to fix the start of the line – notice that the shape of the cursor changes to a small square. Create a line by using the blue cursor keys, [▶] and [▲].

Move the end of the line to the point with coordinates X=3 and Y=1. Press [ENTER] to fix the end of the line.

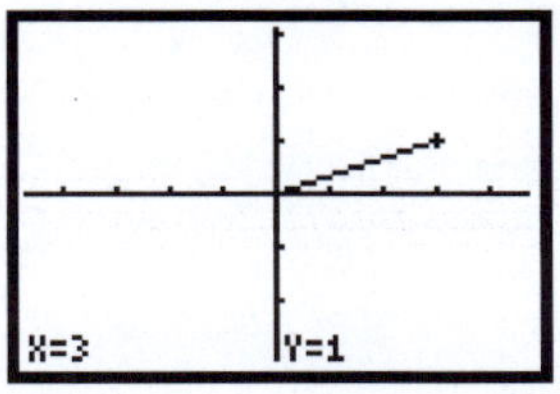

(d) Move the cursor away to another point, press [ENTER] to fix the start of a new line, move the cursor and complete the line.

Experiment by producing several lines. If you make a mistake and want to start again, use **ClrDraw** from the DRAW menu.

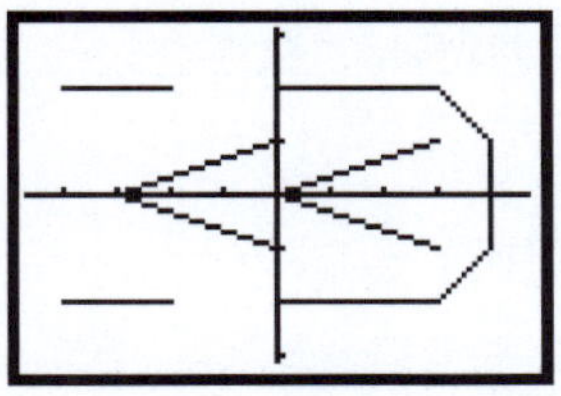

(2) Lined up on the grid

Choose another set of axes, either by pressing [WINDOW] and entering the settings shown here, or by using the program **SETSCRN** (see page 24 of **Foundations Plus**).

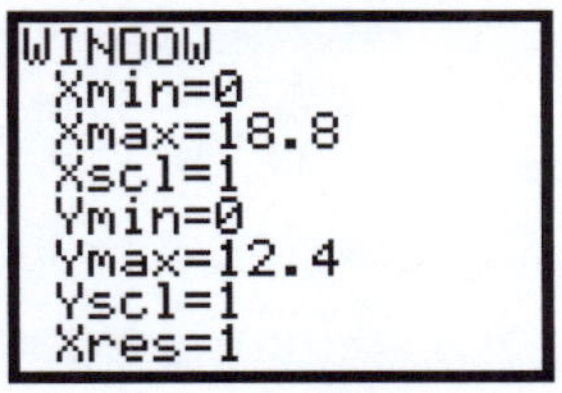

To add a grid to the axes by choosing **GridOn** in the FORMAT menu, press

[2nd] [FORMAT] [▼] [▼] [▶] [ENTER] [GRAPH]

Start on the graphing screen and choose **Line** in the DRAW menu. Draw several lines like the ones shown here. Start and and finish each line on one of the grid points and write down the coordinates of each starting and finishing point.

For example, for the lines shown here, the points were:

(0,0) and (10,6)   (0,0) and (6,10)

(5,4) and (12,11)   (9,3) and (17,10).

(3) A bit steep

(a) Look at the four lines shown above in Activity 2. Which looks the steepest and which the least steep?

Can you place them in order of relative steepness simply by looking at the lines?

(b) Consider the lines on your graphing screen. Decide on their order of relative steepness and make a note beside their starting and finishing points that you recorded in Activity 2.

(4) Grading steepness

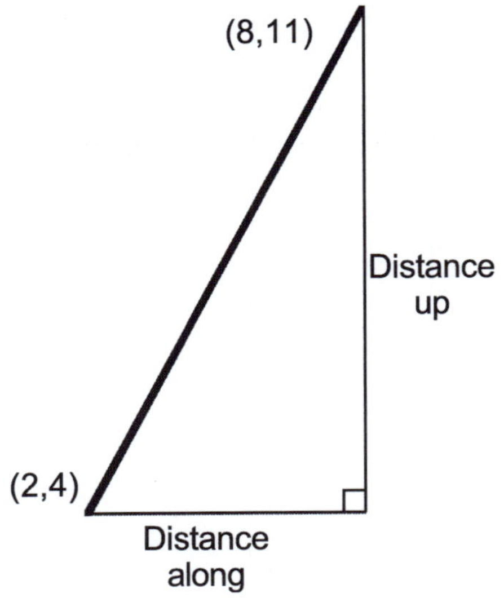

It is useful to be able to measure the steepness of a line numerically.

This number is known as the *gradient* and can be defined as:

$$\frac{\text{Distance up}}{\text{Distance along}}$$

So, for example, a line joining the points (2,4) and (8,11) goes:

11 – 4 = 7 units up

and 8 – 2 = 6 units along.

So its gradient is 7/6 or 1.17 (to 2 d.p.).

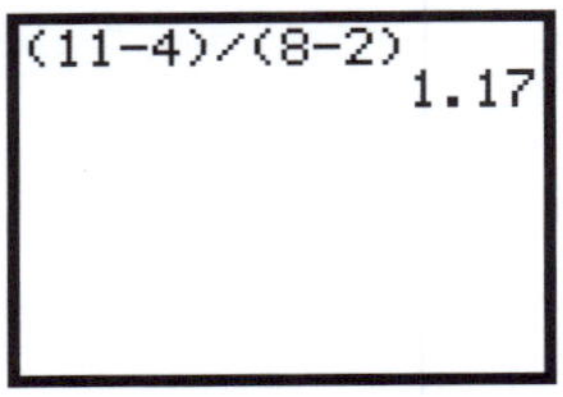

The gradient can be calculated on the calculator's home screen as shown here. Notice that the calculator has been set to show 2 decimal places (see **Foundations Plus**, page 9).

(a) Produce this calculation on your calculator.

(b) Imagine the top point being changed from (8,11) to (10,11).

Will the line be more or less steep?

Will the gradient be bigger or smaller?

Repeat the calculation of the gradient on the home screen. (Note: you can save yourself some time by pressing [2nd] [ENTRY] and editing the previous entry.)

(5) Gradient calculations

Use your calculator to find the gradients of the four lines you drew on your graphing screen in Activity 2.

In Activity 3 you estimated the relative steepness of the four lines. Do the gradients you have calculated support your estimates?

(6) Gradients in general

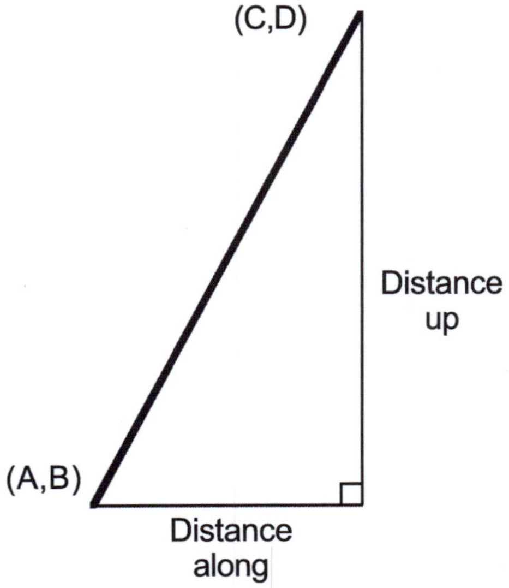

You have now done several specific calculations of gradients. For example, for the line joining (2,4) to (8,11):

$$\text{gradient} = \frac{(11-4)}{(8-2)}$$

Now suppose the end points of the line are the generalised points (A,B) and (C,D).

Write down an expression for the gradient of this line.

(7) Ups and downs

Clear the drawings from your graphing screen by pressing [2nd] [DRAW] **1** [ENTER].

A line can be drawn indirectly by entering a command of the form **Line(A,B,C,D)** on the home screen. Return to the home screen of your calculator by pressing [2nd] [QUIT].

(a) Press [2nd] [DRAW] **2** **5** [,] **5** [,] **10** [,] **8** [)]

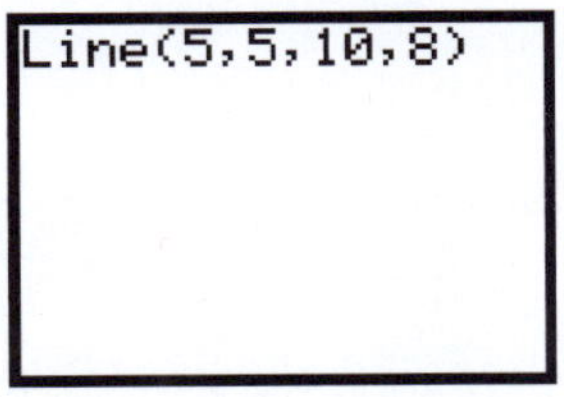

Before you press [ENTER] guess what the line on the graphing screen will look like. Roughly how long will it be? How steep is it? Try to see it in your mind's eye and then press [ENTER] to check your guess.

(b) Guess what each of the following commands will produce and then enter each expression in turn to check your guess. In each case think clearly about the gradient of the line.

Line(9,1,14,4) **Line(5,5,14,4)**

Line(9,1,10,8) **Line(10,8,14,4)**

(c) Enter the appropriate **Line** command to complete the shape.

Notice that this line runs downwards as you move from left to right. Use the formula from Activity 6, gradient $=\frac{(D-B)}{(C-A)}$, to calculate the gradient of this line.

What do you notice?

(8) Are you being negative?

Use the **ZStandard** window settings for this activity.

Which of the following pairs of points are joined by lines with a negative gradient? Guess and then check, either by drawing the line on the graphing screen or by doing an appropriate calculation.

(a) (6,6) and (1,1)
(b) (10,6) and (6,10)
(c) (8, 6) and ($^{-}1$, 6)
(d) ($^{-}1$,$^{-}1$) and ($^{-}6$,$^{-}6$)
(e) ($^{-}5$,0) and ($^{-}6$,$^{-}6$)
(f) ($^{-}6$,$^{-}6$) and (0,$^{-}6$)

(9) A gradient program

Choose the **ZDecimal** window settings again, still leaving the **GridOn** setting selected in the FORMAT menu.

(a) This short program, **GRADI** will draw a line and show its gradient.

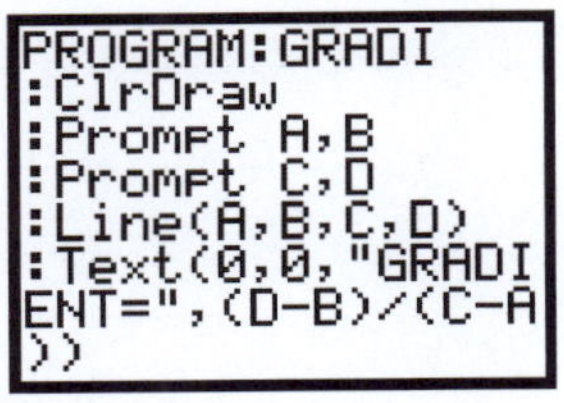

When the program is run you are prompted to enter coordinates for the two points (A,B) and (C,D).

Notice that, incorporated in the final line, is the general formula for a gradient, $\frac{(D-B)}{(C-A)}$.

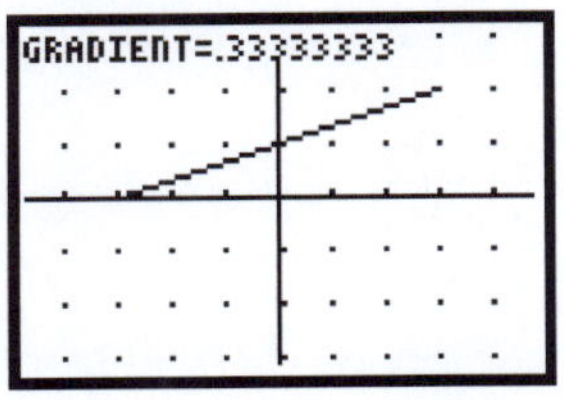

Enter the program in your calculator. The key presses are listed in the solutions if you need to use them.

(b) Execute the program several times, using it to produce lines of different gradients.

(10) Turning the tables

In Activity 9 the coordinates A, B, C and D were chosen by you and the calculator computed the gradients. Now look at this revised form of the program, where the calculator chooses the coordinates using its random number generator.

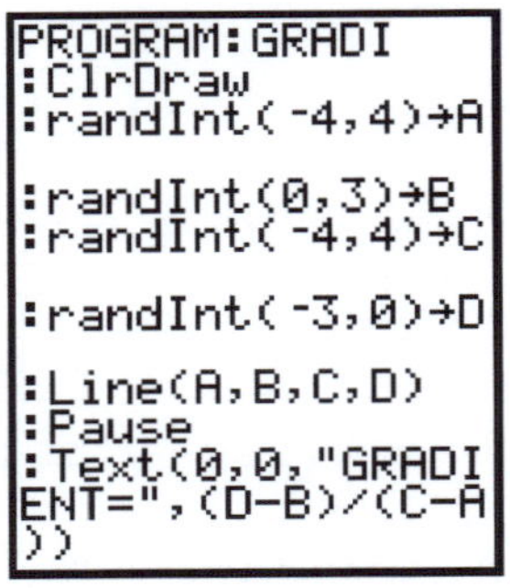

The command **:randInt(-4,4)→A**, for example, tells the calculator to store a randomly chosen integer between $^{-}4$ and 4 in store location A.

When the program is being executed, the extra command **:Pause** causes the calculator to wait before displaying the gradient – this gives you the opportunity to work out what you think the gradient will be. Once you have an answer worked out you can press [ENTER] to check if you were right.

Change the program **GRADI** into the form shown here. Again the necessary key presses are given in the solutions.

Then use the program to give yourself lots of practice at calculating gradients.

(11) Straight line functions

(a) Press [Y=] and enter the function **Y1=(3X+1)/5**.

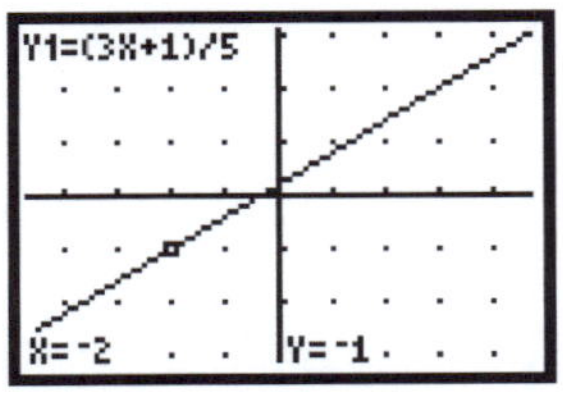

Use [TRACE] to find the coordinates of any two points on the line. Using these coordinates, calculate the gradient of the line. Would you get the same answer if you chose two other points on the line?

Multiply out the brackets from **Y1=(3X+1)/5** and explain how the gradient is related to the resulting equation.

(b) Calculate the gradients of the following straight lines and also multiply out the brackets.

Y2=(2X–1)/4	**Y3=2(X/3 +1)**	**Y4=4(1–X)**
Y5=2	**Y6=$^{-}$2(X–1)**	

(c) In general, what is the gradient of the straight line with equation *Y=mX+c* ?

So finally ...

◊ *If two lines have gradients of 2 and $^{-}$2, is one steeper than the other?*

◊ *If a line joins the origin to the point (S,T), what is its gradient?*

◊ *What are the gradients of horizontal lines such as Y=4 and vertical lines such as X=3?*

You will find some comments and answers to these questions at the back of the book.

Unit 2 A close look at curves

The previous unit looked at the gradient, or steepness, of straight lines. In Units 2 and 3 you will consider a rather harder problem – finding the gradient of a curve.

Be sure to work carefully through Activity 3 as you will be using these ideas again in later units.

In this unit you will be using a member of a family of curves known as cubics. All members of this family have equations that include a term where the highest power of x is 3, that is, x cubed.

(1) Think cubic

Start by resetting the defaults on your calculator. This is explained in **Calculator Maths: Foundations Plus**. Here, among other things, it will ensure that grid points no longer appear on the graphing screen.

Then press [Y=], clear any functions that are displayed and enter the cubic function, **Y1=⁻.15X^3+.1X^2+1.6X**.

Draw the graph using the **ZDecimal** window settings.

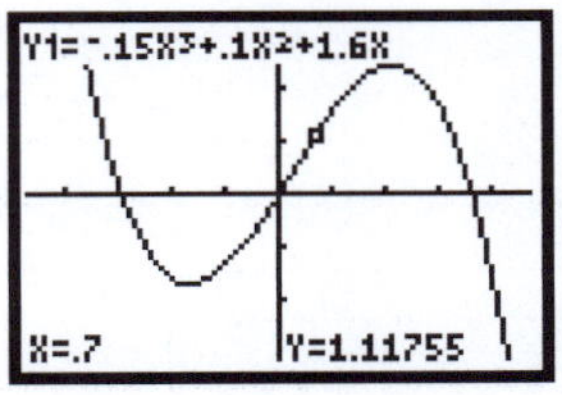

Press [TRACE] and use [◄] and [►] to move along the points that make up the curve. Try to think out for yourself how the calculator has been programmed to produce this graph, before reading on to Activity 2 where it is explained.

(2) Points on the curve

In order to draw the graph of any equation, the calculator first has to work out the coordinates of lots of points. Specifically, it takes the X coordinate of every pixel on the X axis and calculates the corresponding Y coordinate of the point on the graph, using the function defined on the Y= screen. With **ZDecimal** window settings, the extreme left-hand pixels have X coordinates ⁻4.7, ⁻4.6, ⁻4.5, etc. The calculator works out the corresponding Y coordinates by applying Y1 to these values.

You can make it carry out these calculations on the home screen by pressing, for example:

[VARS] [►] **1 1** [(] [(-)] **4.7** [)] [ENTER]

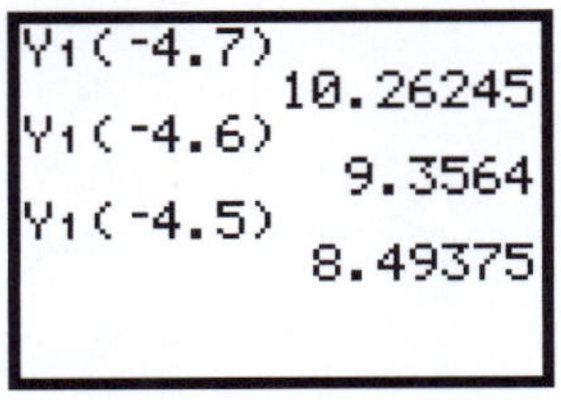

Compare the values of Y1 shown here with those produced using [TRACE]. As you move to the left of X= ⁻3.7 the cursor moves off the screen, but the Y1 values still appear.

Brain stretcher — **Different modes and Xres**

Investigate the effect of changing from **Connected** to **Dot** on the MODE menu, and also changing the **Xres** window setting to its maximum value, which is 8. How does this relate to Activity 2?

(3) How does it do that?

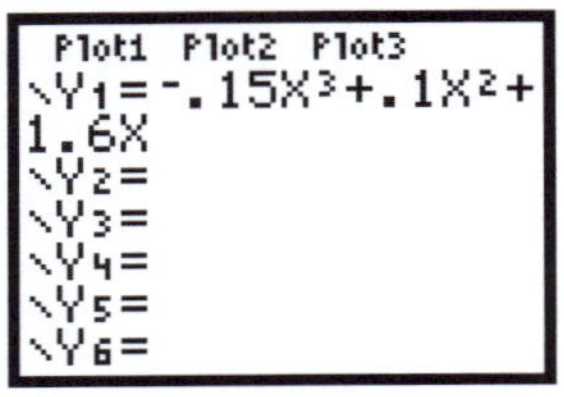

Pretend for a moment that your calculator is broken so that it cannot draw graphs of functions. Press [Y=] and ensure that the function opposite is still entered in Y1. Deselect it by moving the flashing cursor over the = sign and pressing [ENTER]. Now there will be no graph displayed on the graphing screen.

What if you were unable to select it again? Could you draw it another way?

> The symbol Δ is pronounced 'delta' and is the equivalent of capital D in the Greek alphabet. In mathematics it is often used to stand for a small change in something.

Well, you could use a statistical plot – in particular, a line graph. To do this you will need the coordinates of lots of points in two lists. You could use the **seq** command (see **Foundations Plus**, p.13) to produce the X coordinates of the pixels on the graphing screen and then calculate the corresponding Y coordinates using equation Y1.

The X coordinates of the graphing screen are defined by the window settings and extend from **Xmin** to **Xmax** in steps of **ΔX**.

So the appropriate command to store the X coordinates in L1 is
seq(X,X,Xmin,Xmax, ΔX)→L1.

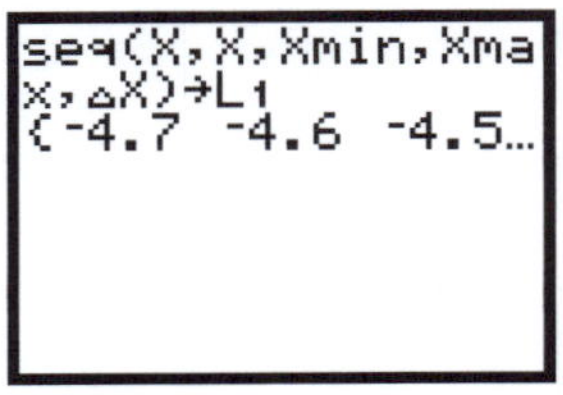

Use these key presses to enter this command now:

[2nd] [LIST] [▶] **5** [X,T,Θ,n] [,] [X,T,Θ,n] [,] [VARS] **1 1** [,]
[VARS] **1 2** [,] [VARS] **1 8** [)] [STO▶] [2nd] [L1] [ENTER]

Now calculate all the corresponding Y coordinates and store them in L2.

[VARS] [▶] **1 1** [(] [2nd] [L1] [)] [STO▶] [2nd] [L2] [ENTER]

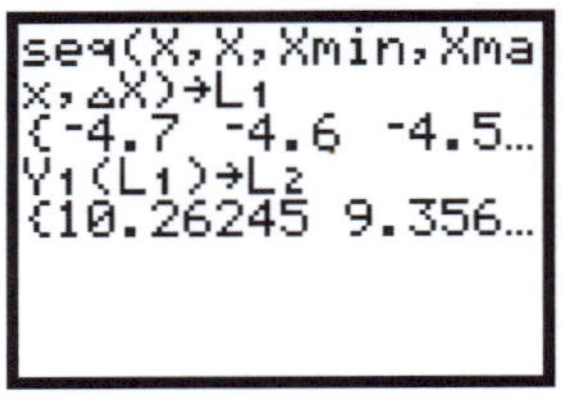

Finally, set up a statistical plot to show the values in L1 plotted against the values in L2. The necessary settings on the Stat Plot menu are shown here.

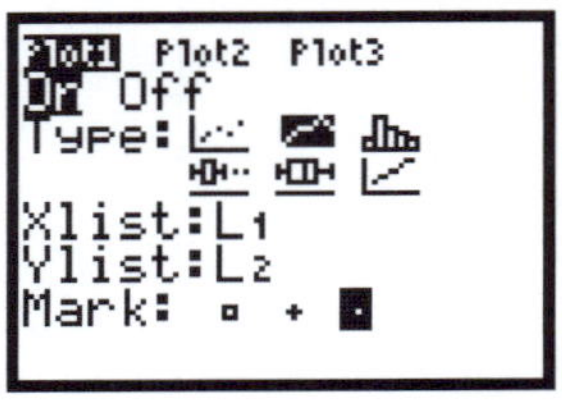

Press [GRAPH] and, hey presto, you have reconstructed the graph of the function.

What if you used a scatterplot rather than a line graph? Try it!

(4) Gradients of the cubic curve

Now concentrate on the gradient of the curve of the cubic **Y1=⁻.15X^3+.1X^2+1.6X**. First go to the Y= screen and select the function again by placing a highlight over the = sign. Also switch off the statistical plot.

In Unit 1 you looked at straight lines and, of course, their gradients are the same right along their length. However, curves are different: their steepness varies as you move along the curve. Use [TRACE] to help you answer these questions: you are not being asked to calculate the gradients, but rather to estimate them by eye.

(a) Is the curve steeper at X=1.5 then it is at X=0?

(b) Where is the gradient of the graph zero? What are the values of X there?

(c) For which parts of the graph is the gradient positive?

(d) For which parts of the graph is the gradient negative?

(e) Estimate whether the gradient at X=0 is greater than or less than 2.
(Hint: Enter as Y2 the equation of a straight line with gradient 2.)

(f) Estimate for what values of X the curve has a gradient of ⁻2.

(g) Where on the screen is the curve steepest?

(h) Where on the screen does the curve have the maximum gradient?

(5) The top of the hill

Ensure that only the equation of the cubic is selected on the Y= screen.

Using [TRACE] move along to the top of the hill where X=2.1.

Option 2 in the ZOOM menu, **Zoom In**, allows you get a closer view of the top of the hill.

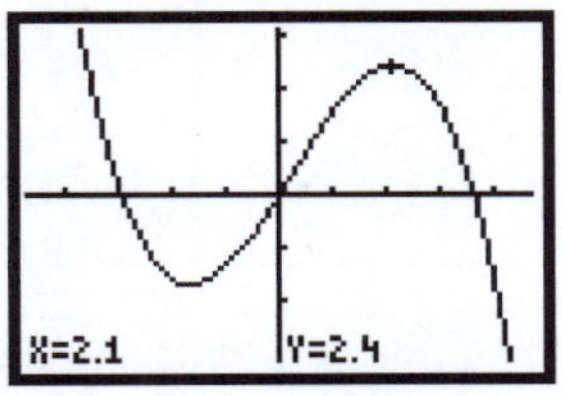

Press [ZOOM] **2**.
Notice the changed shape of the cursor and check that it is still at X=2.1.

Press [ENTER] and [ENTER] again to see closer and closer views of the hilltop.

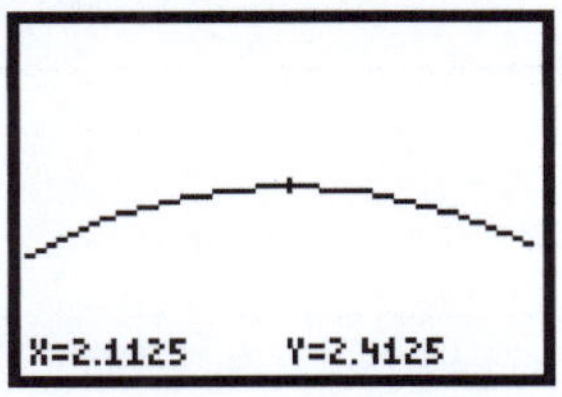

Use the blue cursor keys to move back to what seems to be the very top of the hill and zoom in further by pressing [ENTER].

Repeat this several times. Eventually, what started as a curve seems to be a straight line.

(a) Use [TRACE] to move the cursor along the flat hilltop, where the gradient is zero. Look carefully at the Y coordinates and pick out the largest value of Y that you can find.

(b) Return to the **ZDecimal** window settings and then zoom in on the bottom of the valley, where there is also a gradient of zero.

Pick out the lowest value of Y that you can find.

(6) Halfway up

(a) Start with the **ZDecimal** window settings and use [TRACE] to move to the point where X=1.

Use the method of Activity 5 to zoom in on the curve until it resembles a straight line.

(b) Choose another point of the cubic's curve and zoom in repeatedly. Are there any points on the curve where zooming in does not produce a straight line?

(7) A confusion of gradients

The screen displays below show what you get when you zoom in on $\mathbf{Y1={}^{-}.15X^3+.1X^2+1.6X}$ at the points with X coordinates, -3, -2, -1, 0, 1, 2, 3 and 4. However the order is confused. Can you sort them out?

(a)

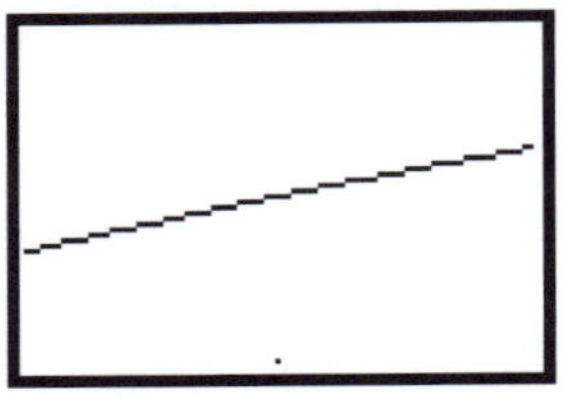

(b)

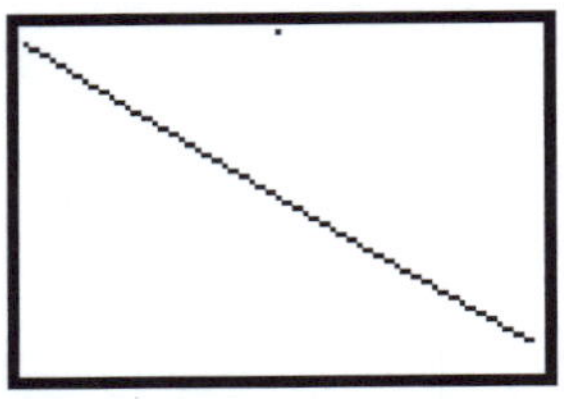

(c)

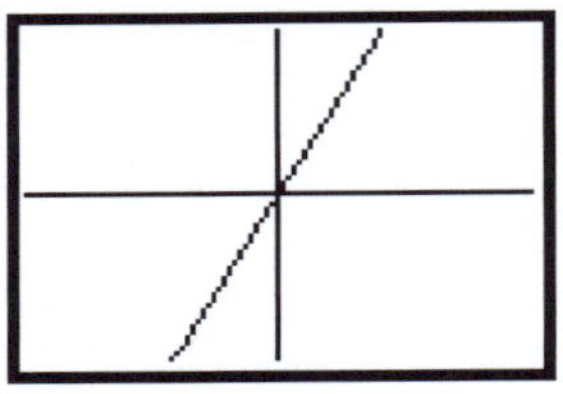

(d)

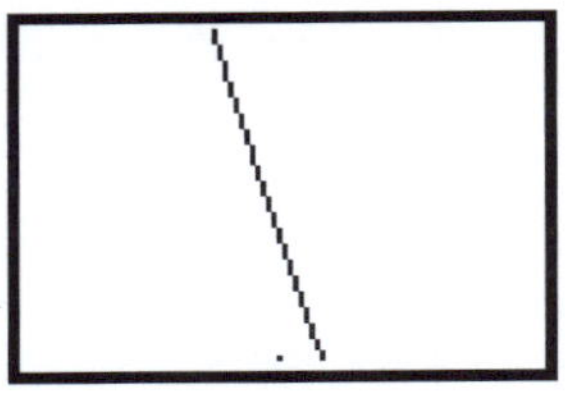

(e)

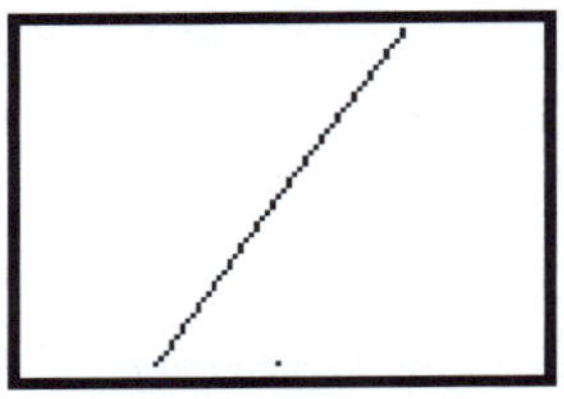

(f)

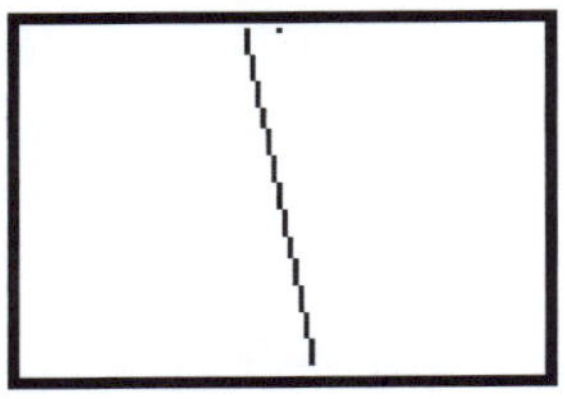

(g)

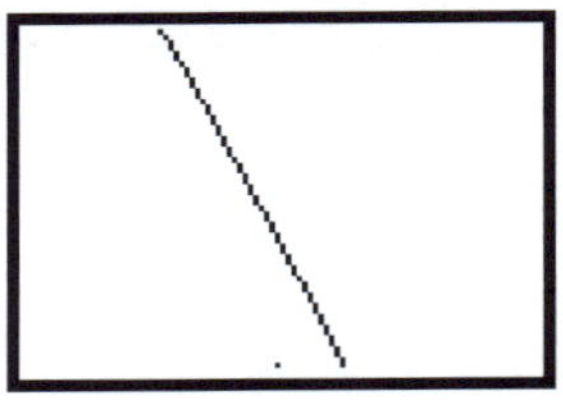

(h)

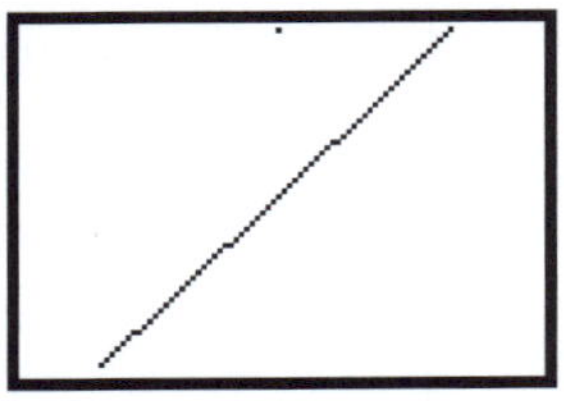

(8) Concentrating on gradients

Below are some functions that have interesting graphs.

For each one:

Enter the function on the Y= screen.

Guess what its graph will look like – that is, form a rough picture in your mind's eye.

Don't leave this step out. Go on ... force yourself!

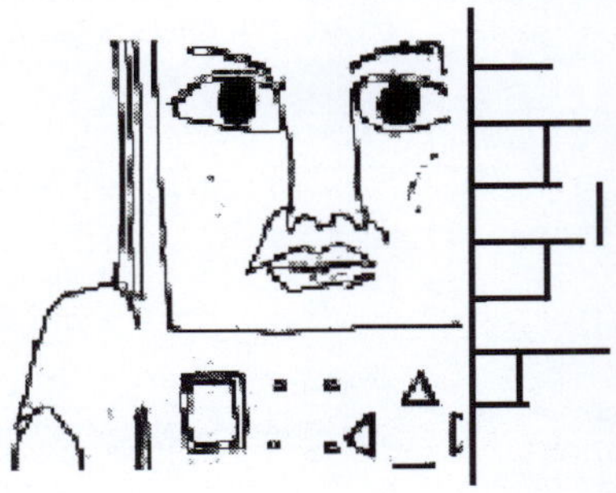

Draw the graph using the **ZDecimal** window settings and check your mind picture.

Describe the graph in terms of its changing gradient.

Imagine zooming in on the graph. Are there any points where the curve would not resemble a straight line?

(a) Y= 1/ (2 – X)

(b) Y= 1/ (1 + X)

(c) Y= 1.5^ X – 3

(d) Y = (X – 1) X (X + 1)

(e) Y = (X – 2) (X – 1) X (X + 1)

(f) Y = (X – 2) (X – 1) X (X + 1) (X + 2)

(g) Y = sin(X)

(h) Y = sin(2X)

(i) Y = tan(X)

(j) Y = $^3\sqrt{}$(X)

So finally ...

Check that you can now answer all these questions – there are brief answers and some comments at the back of the book.

◊ *How does the calculator draw the graph of a function? What calculations must it carry out first?*

◊ *Activity 5 referred to "hills" and "valleys" which have gradients of zero. Are there any curves that have points with gradients of zero, but which are not "hills" or "valleys"?*

◊ *A curve is made up of lots of tiny straight lines. Is this true or false?*

◊ *What does the calculator command **ΔX** do?*

◊ *Why do you think zooming in on a curve eventually produces a straight line?*

Unit 3 Gradient of a curve at a point

In Unit 1 you saw how to calculate the gradient of a straight line joining two points (A,B) and (C,D) using the formula:

$$\text{gradient} = \frac{(D-B)}{(C-A)}$$

After studying Unit 2 you will have realised that a curve has different gradients at different points and that the more you zoom in on it, the more closely the curve resembles a straight line.

In this unit these ideas will be used to calculate the gradient of a curve *at a particular point.*

(1) Gradient of a line joining two points on the cubic

If necessary, re-enter on the Y= screen the function

$$\mathbf{Y1={}^{-}.15X^3+.1X^2+1.6X}$$

Deselect or clear any other functions.

Press [ZOOM] **4** to select the **ZDecimal** window settings.

(a) Press [TRACE] and move to the point on the curve with X coordinate equal to 1.

Write down the value of the Y coordinate at this first point.

(b) Using [▶] move to the point on the curve with X coordinate equal to 2. Write down the coordinates of the second point.

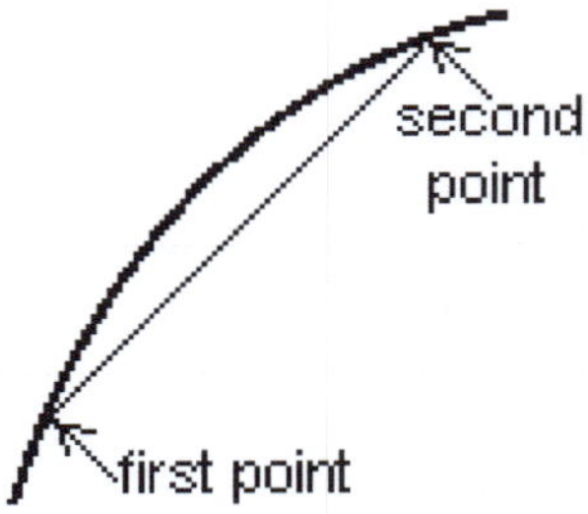

(c) On the home screen, calculate the gradient of the line joining these two close points using the formula

$$\text{gradient} = \frac{(D-B)}{(C-A)}$$

Notice that this is *not* the gradient at any point on the curve but *is* the gradient of the line joining two points on the curve. You could get a closer approximation to the gradient *at* the first point if you moved the second point closer to it.

However, you can save yourself some tedious work by automating the process.

(2) Automating the calculation

You will be doing a lot of calculations like the previous one, so here is a way to speed things up a bit.

(a) Firstly you can store the coordinates of the point where the gradient is needed in store locations A and B.

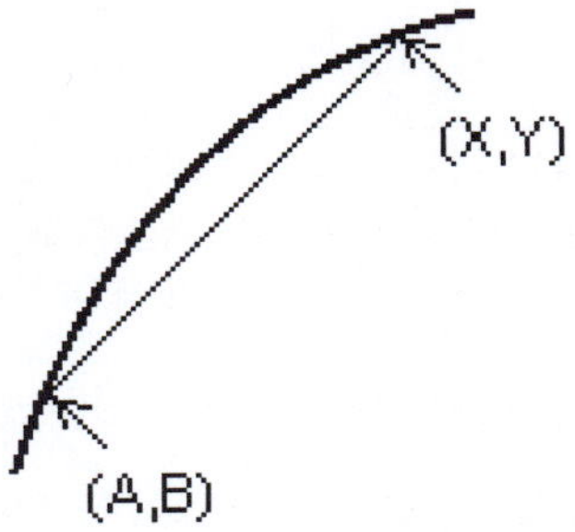

To do this press [TRACE] and move to the point (here it is where X=1).

Move to the home screen and store the current values of X and Y in A and B by entering

X→ A: Y→ B

You need to press [X,T,Θ,n] [STO▸] [ALPHA] **A** [ALPHA] **:**
[ALPHA] **Y** [STO▸] [ALPHA] **B** [ENTER]

(b) Now press [TRACE] and use [▸] to move to the second point (here it is X=2).

(c) On the home screen enter **(Y–B)/(X–A)**.

You should get the same answer as in Activity 1.

(d) Press [TRACE] and move one pixel to the left (to X=1.9).

Press [2nd] [QUIT] to return to the home screen.

You can repeat the last calculation simply by pressing [ENTER].

This is the gradient of the line joining (A,B) to the point on the curve where X=1.9.

(e) Repeat part (d) several times. With each press of [TRACE] [◂] [2nd] [QUIT] [ENTER] you are calculating a new gradient joining (A,B) to nearer points on the curve.

The first four gradients are shown here.

Continue with this process, repeating the calculation until the second point (X,Y) is as close as possible to (A,B). Investigate what happens when the two points are on top of each other.

(f) Compare the gradient values that you get on the home screen with those you might expect by looking at the graph. As (X,Y) gets closer to (A,B) would you expect the gradient of the line joining the points to be increasing or decreasing?

(3) Moving (X,Y) away to the left

(a) Now suppose the point (X,Y) is on the left of (A,B) and moves gradually away from (A,B). What would you expect to happen to the gradient of the line joining the points?

Repeat the process of Activity 2(d) in order to check your expectations.

(b) You still do not know the gradient of the curve *at* the point (A,B) but you do have two good estimates:

> the gradient of the line joining the two points when the second point is one pixel to the *right* of (1,1.55). You calculated this in Activity 2.
>
> the gradient of the line joining the two points when the second point is one pixel to the *left* of (1,1.55). You calculated this in this activity.

It is likely that the true gradient at (A,B) lies somewhere between these two estimates. Make a note of them now. They are the best that can be found using this method with the **ZDecimal** window settings because they use the closest points on either side of (1,1.55).

(4) Zooming in for better estimates

As you saw in Unit 2, if you zoom in on a point, the curve increasingly resembles a straight line. It will also enable you to choose a closer point for the calculation of a gradient.

You can do this using a variation of the automated calculation as follows.

(a) Make sure that you are using the **ZDecimal** window settings by pressing [ZOOM] **4**.

Press [TRACE] and move to the point on the curve with X coordinate equal to 1. Press [▶] once more to move 1 pixel to the right.

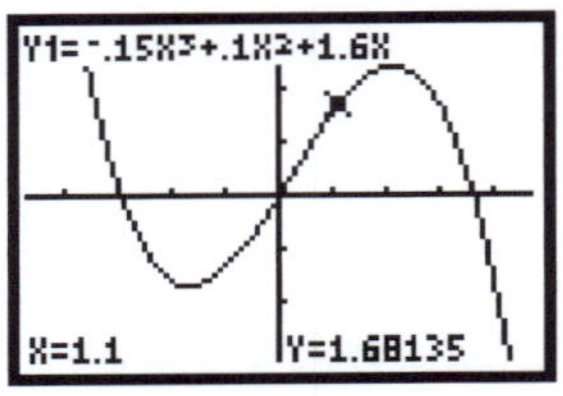

(b) Go to the home screen and enter **(Y–B)/(X–A)**. You will probably be able to do this by pressing [2nd] [ENTRY] [ENTER].

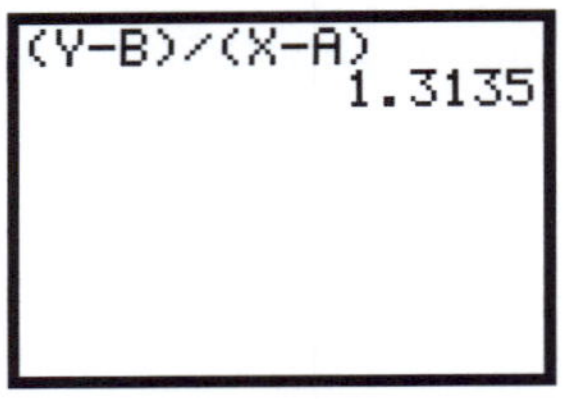

(c) Press [TRACE] [◀] to return the cursor to X=1.

Press [ZOOM] **2** to select **Zoom In** and press [ENTER] to activate it.

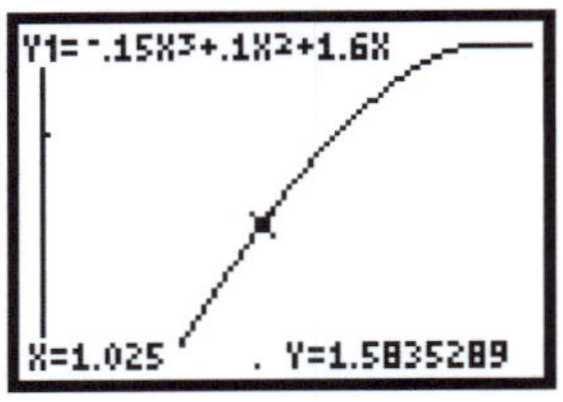

(d) Press [TRACE] [▶] to move the cursor 1 pixel to the right of X=1.

(e) Return to the home screen. Calculate **(Y–B)/(X–A)** again, simply by pressing [ENTER].

```
(Y-B)/(X-A)
              1.3135
          1.34115625
         1.347806641
         1.349452759
```

(f) Repeat steps (c) to (e) several times to produce increasingly better estimates of the gradient at the point of (1,1.55).

(g) Notice how the calculated values appear to be converging to a particular value as the two points get very, very close together.
Repeat the process a few more times and write down where the calculated values appear to be converging towards. This value is known as the *limit*.

(5) Zooming in from below

Repeat Activity 4 to find estimates of the gradient but this time make the second point to the left, just *below* (1,1.55).

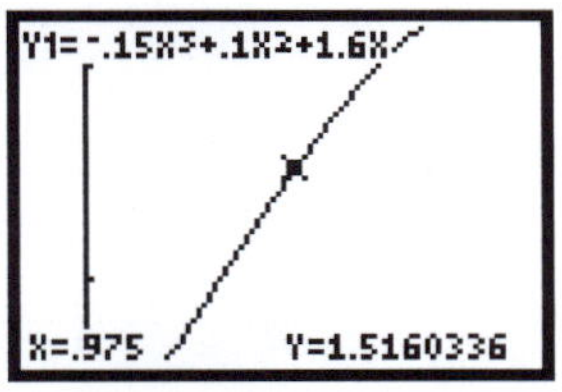


```
(Y-B)/(X-A)
              1.3835
          1.35865625
```

(6) Is it horizontal?

Press [ZOOM] **4** to select the **ZDecimal** window settings once more.

```
X→A:Y→B
            -1.69405
(Y-B)/(X-A)
```

(a) Press [TRACE] and move to the "valley" point on the curve. By looking carefully at the Y values, decide which value of X is nearest the bottom of the valley. Store these values of X and Y in A and B.

(b) By zooming in on the right (as in Activity 4), find the best estimate that you can for the gradient of the curve at the point you identified in part (a).

(c) Repeat parts (a) and (b) to find a point near the top of the "hill" and find the best estimate for the gradient of the curve at this point.

(7) A different curve

Deselect the function Y1 – do not clear it as you will need it again in the next unit.

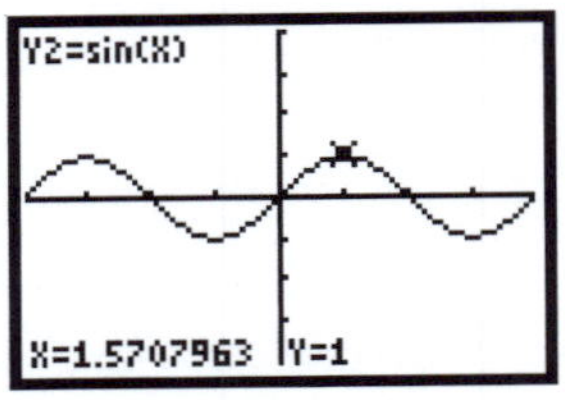

Enter the function **Y2 = sin(X)**.

Make sure that your calculator is operating in **Radian** mode.

Press [ZOOM] **7** to select the **ZTrig** window settings.

(a) Press [TRACE] and investigate the values of X and Y at the various hills and valleys displayed.

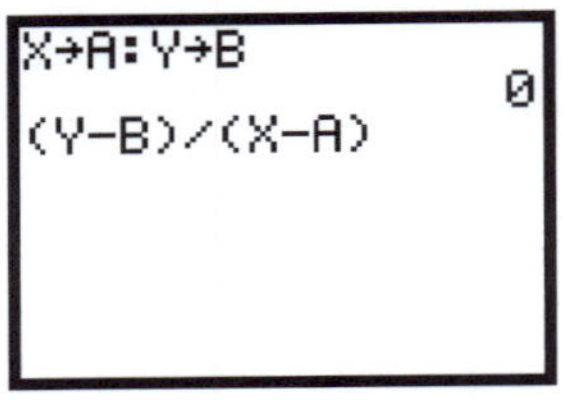

(b) Use the method developed above to get an estimate for the gradient of the curve at the origin. Zoom in and get a series of better estimates.

On the basis of these estimates, what value do you expect the gradient to have *at* the origin?

(c) What do you expect the gradient of the curve to be at the other points where it crosses the X-axis? Use the method developed above to check.

Brain stretcher — **An awkward curve**

Change the function Y2 to make it **Y2 = (X+2)/(X–2)**.
View the curve using the **ZStandard** window setting.
Use Trace to move to a point just left of X=2.
Use **X→A: Y→B**.
Move one pixel to the right and enter **(Y–B)/(X–A)**.
Is the gradient really about 368?
Can you get an estimate for the gradient of the curve at X=2?

So finally ...

◊ *In this unit you have calculated estimates of the gradients of curves at particular points and have found that as you zoom in on the graph the gradient converges towards a particular value.*
How far do you have to zoom in before you can be sure of what that value is?

Unit 4 Tabulating and programming gradients

In the last unit, in order to approximate to the gradient at a particular point you zoomed in repeatedly, calculating the gradient over successively smaller intervals to the right of the point.

Rather than zooming in on the point at which the gradient is to be calculated, it is possible to use the calculator's table facility using values of X that are closer and closer together. This method is more suited for development into a calculator program, and this what is done in the second part of the unit.

(1) Setting the table

Once again concentrate on finding the gradient of the cubic function: **Y1=⁻.15X^3+.1X^2+1.6X**.

If necessary, re-enter the function on the Y= screen and clear any other functions. Choose the **ZDecimal** window settings.

Press [TRACE] and move the cursor to the point where X=1 and Y=1.55.

Now press [2nd] [TBLSET].

Set the screen as shown here. Note that **TblStart** is the x-coordinate of the point of interest, and **ΔTbl** is the step size (or increment).

Press [2nd] [TABLE] and you should see the screen shown here. Notice that the second Y1 value is actually 1.68135: the more accurate value is shown at the bottom of the screen.

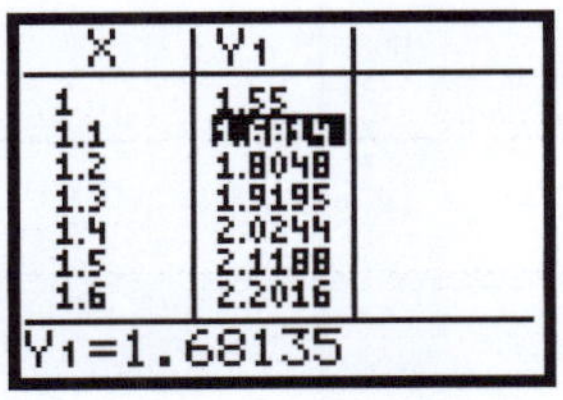

Now you can calculate the gradient between the point of interest (1,1.55) and the next point with coordinates in the table (1.1,1.68135).

The gradient is $\dfrac{(1.68135-1.55)}{(1.1\text{-}1.0)}$.

Go to the home screen and calculate the gradient using these values.

(2) A more general calculation

Notice that the two X-coordinates used in the last calculation could have been entered on your calculator as

X	(the current value is 1) and
X + **ΔTbl**	(the current value is 1 + 0.1 = 1.1).

The corresponding Y-coordinates could have been entered as

Y1(X)	(the current value is 1.55) and
Y1(X + **ΔTbl**)	(the current value is 1.68135).

So the gradient could be calculated as $\frac{Y1(X+\Delta Tbl) - Y1(X)}{\Delta Tbl}$.

Enter this complicated expression on your home screen by pressing the following sequence of keys:

[(] [VARS] [▸] **1 1** [(] [X,T,Θ,n] [+] [VARS] **6 2** [)] [−]
[VARS] [▸] **1 1** [(] [X,T,Θ,n] [)] [)] [÷] [VARS] **6 2**

Check this key sequence carefully before pressing [ENTER].

You should get the same value for the gradient as before.

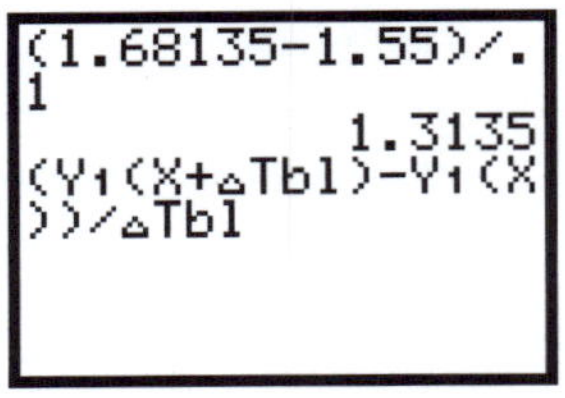

(3) Reducing ΔTbl

(a) Return to the TBLSET menu and decrease **ΔTbl**, making it ten times smaller, i.e. .01.

Look at the second line of values on the Table screen. They are the current values of X + **ΔTbl** and Y1(X + **ΔTbl**).

Return again to the home screen and calculate the gradient again. You can do this simply by pressing [ENTER] to repeat the previous calculation with the current values. Remember that this is the gradient of a line joining the point at (1, 1.55) to another point on the curve very close to it.

(b) Repeat the steps in part (a) several times, using ever smaller values for **ΔTbl** by dividing the previous value by 10.

You should see a series of gradients, apparently converging to a particular value. These are ever better approximations for the gradient of the curve *at* the point where X=1.

(4) Practising the process

Repeat the process used above to calculate a series of good approximations for the gradient of the curve at the point where X= ⁻1.7.

Remember to trace on the curve to this point before you start (this stores the value ⁻1.7 in X).

(5) Programming the process

The process described in the previous activities consists essentially of a repeating loop within which **ΔTbl** is made ten times smaller and the gradient calculated. Because of this, it is easy to write a short program to carry out the procedure. The successive values of **ΔTbl** with corresponding values of the gradient can be stored in lists L1 and L2 and then displayed. This will also be useful in Unit 6, enabling these values to be compared with values calculated using another method.

Below is a possible form of the program.
Use the key presses shown to enter the program into your calculator.

PROGRAM: GRADY	[PRGM] [◄] **1 GRADY** [ENTER]	
:Prompt X	[PRGM] [►] **2** [X,T,Θ,*n*] [ENTER]	Input where the gradient is to be calculated.
:1→ΔTbl	**1** [STO►] [VARS] **6 2** [ENTER]	Set an initial value for ΔTbl.
:For(C,1,10)	[PRGM] **4** [ALPHA] **C** [,] **1** [,] **10** [)] [ENTER]	Set up a loop that will be used for the values of C from 1 to 10.
:ΔTbl/10→ΔTbl	[VARS] **6 2** [÷] **10** [STO►] [VARS] **6 2** [ENTER]	Reduce ΔTbl by a factor of 10.
:ΔTbl→L1(C)	[VARS] **6 2** [STO►] [2nd] [L1] [(] [ALPHA] **C** [)] [ENTER]	Store ΔTbl in L1.
:(Y1(X+ΔTbl)–Y1(X))/ ΔTbl→ L2(C)	[(] [VARS] [►] **1 1** [(] [X,T,Θ,*n*] [+] [VARS] **6 2** [)] [–] [VARS] [►] **1 1** [(] [X,T,Θ,*n*] [)] [)] [÷] [VARS] **6 2** [STO►] [2nd] [L2] [(] [ALPHA] **C** [)] [ENTER]	Calculate the gradient and store in L2.
:End	[PRGM] **7**	Return to the beginning of the loop.

Finally press [2nd] [QUIT] to leave the program editor.

(6) Executing GRADY

(a) Before you first use the program it is important to clear any existing values in lists L1 and L2.

Now execute program **GRADY** and use it to produce a series of approximations for the gradient of the function in Y1 at the point where X=2.

```
ClrList L1,L2
                Done
prgmGRADY
X=?2
                Done
```

Executing GRADY

When the program has run, press [STAT] **1** to see the stored values.

```
L1      |L2      |L3     2
 .1     |.1185   |------
 .01    |.19199  |
 .001   |.1992   |
 1E-4   |.19992  |
 1E-5   |.19999  |
 1E-6   |.2      |
 1E-7   |.2      |
L2(6) =.1999992
```

Notice that as you run the cursor down the column of values in L2, the more accurate values are displayed at the bottom of the screen. For example, as shown here the gradient corresponding to 1E-6 has been rounded to 0.2 in the list, but the actual value stored in L2(6) was 0.1999992.

It seems that the gradient of the curve at X=2 is likely to be 0.2 and you can tell that this is so when Δ**Tbl** = 1E⁻7, or 0.0000001.

(b) Use the program again to investigate the gradient of the curve at points where X=3, X=4, X=5 etc.

In each case write down the gradient and the value of Δ**Tbl** at which you are confident that you know the gradient.

Brain stretcher **Pushing GRADY to the limit**

Look what happens if you change the third command in program **GRADY** so that it goes round the loop 20 times. Does this indicate that the true gradient is zero? Try for various values of X.

```
L1      |L2      |L3     2
 1E-11  |.2      |
 1E-12  |0       |
 1E-13  |0       |
 1E-14  |0       |
 1E-15  |0       |
 1E-16  |0       |
 1E-17  |0       |
L2(11) =.2
```

Can you explain the problem?

(7) The best value for ΔTbl

Your answers to Activity 7 will have indicated that by the time ΔTbl is as small as 1E⁻7, or 0.0000001, the approximation of the gradient value is as good as it is going to get (as far as this calculator is concerned). Also, if you tried the Brain stretcher, you will have found that for even smaller values of X the gradient calculation runs into problems.
This means that if you are only interested in the gradient of the function in Y1 at a particular value of X you could use a simplified form of the program as shown below.

Program listing

```
PROGRAM:GRADIENT
:Prompt X
:1E-7→ΔTbl
:Disp (Y1(X+ΔTbl
)-Y1(X))/ΔTbl
```

The program in use

```
prgmGRADIENT
X=?1
            1.35
            Done
X=?2
              .2
            Done
```

Create the program **GRADIENT** shown here. You can either enter all the commands from scratch or use some of the commands from program **GRADY** as suggested below.

Create a new program called **GRADIENT**.	[PRGM] [◄] **1 GRADIENT** [ENTER]
Recall the commands from **GRADY**.	[2nd] [RCL] [PRGM] [◄] and choose **GRADY** from the menu. [ENTER] [ENTER]
Return to the top of the program	[ALPHA] [▲] [ALPHA] [▲]
Edit the second command.	[▼] [►] [2nd] [INS] [2nd] [EE] [(-)] **7**
Remove the next three commands.	[▼] [CLEAR] [DEL] [CLEAR] [DEL] [CLEAR] [DEL]
Edit the next command to display the gradient.	[2nd] [INS] [PRGM] [►] **3** [▼] [►] [►] [►] [►] [DEL] [DEL] [DEL] [DEL] [DEL]
Delete the last command and return to the home screen.	[DEL] [DEL] [2nd] [QUIT]

(8) Exercise with the program

Use the program **GRADIENT** to calculate and write down values for the gradient of the curve **Y1=⁻.15X³+.1X²+1.6X** at the following values of X.

(a) X = ⁻3 (b) X = ⁻2 (c) X = ⁻1 (d) X = 0 (e) X = 2 (f) X = 3 (g) X = 4

So finally ...

◊ *This unit showed that it is possible to calculate estimates for gradients using* $\frac{Y1(X+\Delta Tbl) - Y1(X)}{\Delta Tbl}$ *with a small value of* ***ΔTbl****, enabling a program to be used.*
How does this compare with the method you used in Unit 3?

Unit 5 Dee why by dee ex

In this unit you will begin to use some of the calculator's built-in facilities for calculating gradients and drawing a tangent.

" Why was I not told about this before?", you cry! The answer is that the aim of this book is that you should end up understanding the fundamental concepts of Calculus. By working through the first four units you should have built up a firm foundation of knowledge about gradients and how the calculator goes about calculating them. But now you should be ready to explore the built-in facilities of the calculator and to use them with understanding.

(1) Introducing delta

In the program **GRADIENT** in Unit 4 you used the formula:

$$\text{gradient} = \frac{\text{Y1(X+ }\Delta\text{Tbl)} - \text{Y1(X)}}{\Delta\text{Tbl}}$$

However, there was no real need for the small change in the x-coordinate to be stored in **ΔTbl**. Any store location would have served equally well for the purposes of the program.

δx is the mathematical notation that is frequently used to stand for a small change in the x-coordinate. δ is the Greek lower-case letter called 'delta', the equivalent to the English letter d. The equivalent capital letter is Δ, as in **ΔTbl**.

So the formula above could well be expressed as:

$$\text{gradient} = \frac{\text{Y1(X+ }\delta x) - \text{Y1(X)}}{\delta x}$$

Just as δx is used to stand for a small change in the x-coordinate so δy is used to stand for the corresponding small change in the y-coordinate and the gradient is then calculated using the formula:

$$\text{gradient} = \frac{\delta y}{\delta x}$$

Are you convinced that Y1(X+δx) – Y1(X) does indeed represent a change in the *y*-coordinate?

Write down a clear explanation of why.

(2) CALC option 6

Because finding the gradient of a function at a particular value of *x* is required fairly frequently, the calculator has a pre-programmed facility to do just this. It is option 6 in the CALC menu and is labelled **dy/dx**.

This is pronounced "*dy* by *dx*".

The symbol $\frac{dy}{dx}$ is used in mathematics to stand for the derivative of y with respect to x, or the gradient of the curve at a particular point.

It is defined as the limit of $\frac{\delta y}{\delta x}$ as δx approaches 0.

Make sure **Y1=⁻.15X³+.1X²+1.6X** is selected on the Y= screen.

Select the **ZDecimal** window settings and press

[2nd] [CALC] **6**

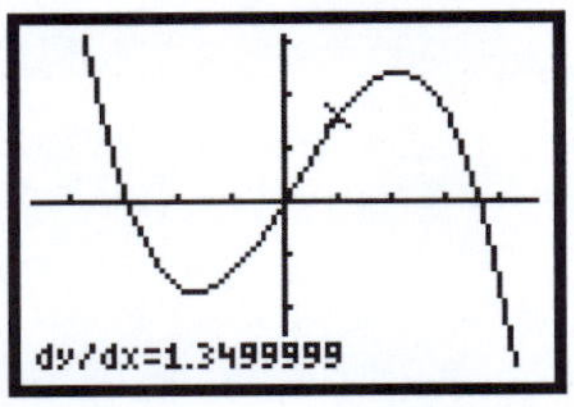

Use the [▶] key to move along the curve, just as if you were using Trace.

When the cursor is highlighting the point with coordinates X=1 and Y=1.55 press [ENTER] and you will see the calculated value of **dy/dx** displayed on the screen.

It is important to realise that the figure given by the calculator as *equal to dy/dx* is, in fact, just another approximation to the gradient of the line at X=1. It has actually been calculated by considering the gradient of a line joining two points on the curve that are very close together, using a similar program to **GRADIENT** which you used in Unit 4. According to the calculator manual, the difference between the x-coordinates of the two points used for the calculation (i.e. the value of δx) is 10^{-5}.

(3) Calculating with CALC

Use option 6 in the CALC menu to find approximations for the gradient of the curve **Y1=-.15X³+.1X²+1.6X** at the following values of X.

(a) X = ⁻3 (b) X = ⁻2 (c) X = ⁻1 (d) X = 0 (e) X = 2 (f) X = 3 (g) X = 4

HINT: When using **dy/dx** *you can save time by entering the required X value rather than using the cursor keys to move along the line, so in part (a) try pressing*

[2nd] [CALC] **6 ⁻3** [ENTER].

(4) Hills and valleys

Using the **ZDecimal** window settings, at what values of X (correct to 1 d.p.) is **dy/dx** closest to 0? These are the two points where the gradient is zero, so if you zoomed in on the graph at that point you would get a horizontal line.

These points are sometimes called hills and valley, or peaks and troughs – but the proper mathematical terms are the *local maximum and local minimum points.*

Why do you think the adjective *local* is used?

(5) A different curve

In Unit 3 you looked at the gradient of the sine curve. Enter the function **Y2 = sin(X)** and make sure that your calculator is operating in Radian mode.

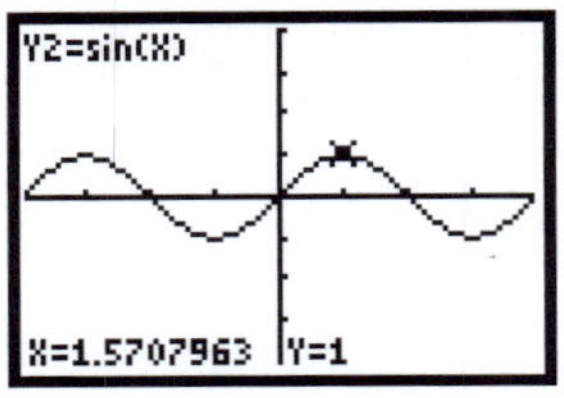

Press [ZOOM] **7** to select the **ZTrig** window settings.

Use **dy/dx** to investigate the gradient of this curve. What is the maximum gradient that you can find?

(6) A discontinuous curve

If you tried the Brain stretcher: *An awkward curve* in Unit 3 you will have already met the function **Y2 = (X+2)/(X–2)**.

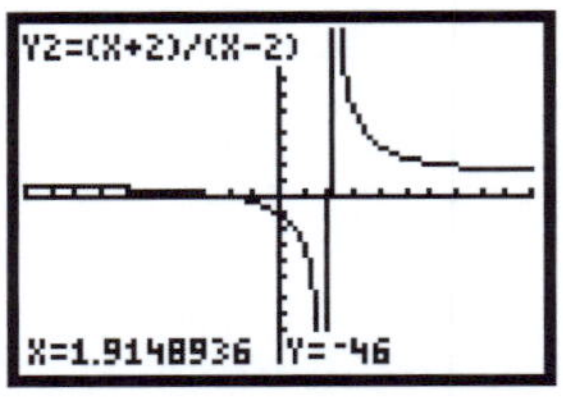

Enter this function now and look at its curve using the **ZStandard** window settings. The graph seems to suggest that at X=2 the curve cuts the x-axis, i.e. at X=2, Y=0.

Try using **dy/dx** at X=2. Problems! No value appears.

This curve looks very different using the **ZDecimal** window settings. Again try calculating **dy/dx** at X=2.

Now use the calculator's Table function and look for the value of Y2 at X=2.

It becomes clear that there is *no* value of Y2 when X=2. For values of X slightly less than 2, Y2 has very large negative values, while for values of X slightly more than 2, Y2 has very large positive values. But at X=2 there is no value of Y2, no point on the graph and therefore no gradient at that point.

The curve is said to have a *discontinuity* at X=2.

Brain stretcher **Try to be awkward**

Can you produce more curves with discontinuities like the ones below?

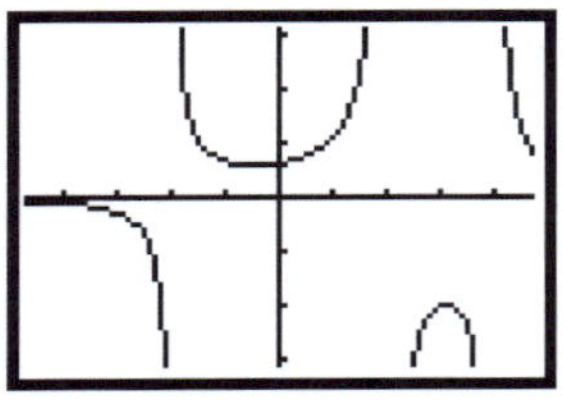

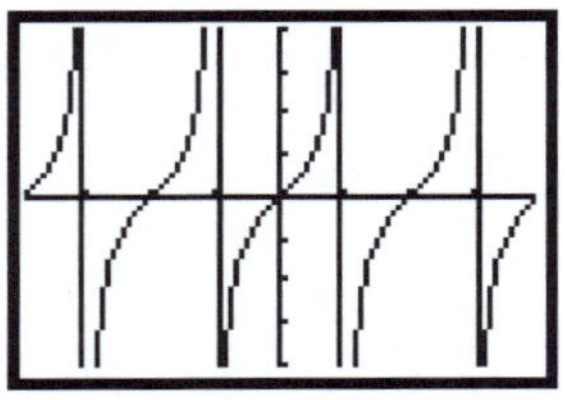

(7) Drawing tangents indirectly

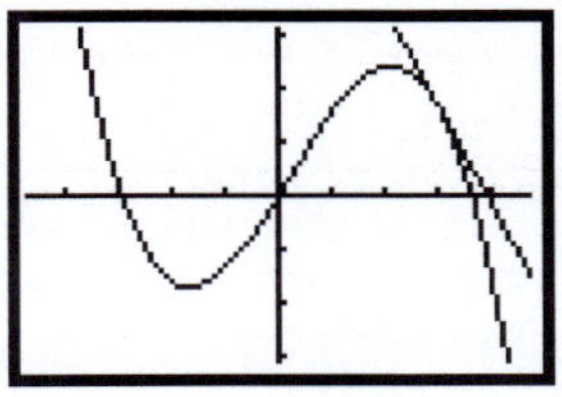

The tangent of a curve at a point is defined as the straight line passing through that point which has the same gradient as the curve at that point. It can be thought of as the line that just touches the curve at the point. To the right you can see a tangent drawn to the cubic curve at the point where X=3.

The calculator has a command that allows you to draw tangents to curves. It is option 5 in the DRAW menu, **Tangent**. As you saw in Unit 1, many of the drawing commands can be used either *directly* on the graphing screen or *indirectly* on the home screen, and this is true for the command **Tangent.**

Make certain that only **Y1=-.15X^3+.1X^2+1.6X** is selected on the Y= screen, and plot it using the **ZDecimal** window settings.

Move to the home screen and enter **Tangent(Y1,3)** by pressing
[2nd] [DRAW] **5** [VARS] [▸] **1 1** [,] **3** [)] [ENTER]

You should see the tangent to the curve, as shown above. You may notice that there is a slight pause before the drawing takes place. This is because the calculator has to work out the gradient of the tangent, *dy/dx*, in order to draw the graph.

(8) Drawing tangents directly

Now use **Tangent** to draw directly on the graphing screen, perhaps to draw the tangent at X=$^-$3.

With the graphing screen displayed, press [2nd] [DRAW] **5** and either move the cursor along the curve to $^-$3 or, better, enter $^-$3 on the keyboard.

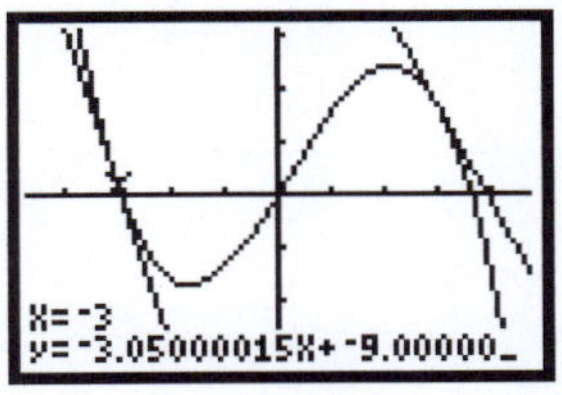

Press [ENTER] and you should see the display on the right. Notice that this time the equation of the tangent line is displayed at the bottom of the screen.

Look back at your answer to Activity 3(a). Can you see how the calculator has calculated the gradient of the curve at $^-$3 and used this as the gradient for the tangent?

Add some more tangents to the curve where X=$^-$2, X=$^-$1 etc.

(9) How does it calculate the equation of the tangent?

Because the tangent is a straight line, its equation is of the form Y = *m*X + *c*.

You have seen how the calculator determines the value of *m* – it is the same as the gradient of the curve, **dy/dx**. How does it work out the value of *c*?

Suppose the point on the curve has coordinates (A,B). This point lies also on the tangent line so (A,B) must fit the equation:

$$Y = mX + c.$$

Substituting A for X and B for Y gives $B = mA + c$

and re-arranging gives $c = B - mA$.

Because A, B and *m* are all known, *c* can be calculated.

(10) Draw the tangent yourself

(a) Draw the graph of **Y1=-.15X³+.1X²+1.6X** using the **ZDecimal** window settings.

Use **dy/dx** to calculate the value of the gradient at X=2.5 and press [TRACE] to calculate the value of Y. Write down these values correct to 4 decimal places.

(b) To draw the tangent you will need to use an equation of the form Y = *m*X + *c*. The point on the curve is (A,B), and so *c* = B – *m*A. Calculate the value of *c*.

(c) Enter the equation of the tangent on the Y= screen and display the straight line. Check that it just touches the curve at X=2.5.

(d) Now deselect the tangent equation on the Y= screen and draw it another way: you can use **Tangent** from the DRAW menu. Draw the tangent at the same point and check that this drawing matches the one drawn in part (c).

(e) Repeat parts (a) to (d) to draw a tangent at a different point on the curve.

(11) Tangential screensnap

Here is a short program that can be used to create patterns like those shown below.

```
PROGRAM:LOTATANS
:For(C,Xmin,Xmax
,5ΔX)
:Tangent(Y2,C)
:End
```

Enter the program on your calculator. If you need help, the key presses are given in the solution.

To use the program you must have an appropriate equation entered in Y2. What equations have been used in Y2 for the diagrams below?

Can you recreate these patterns or similar ones of your own?

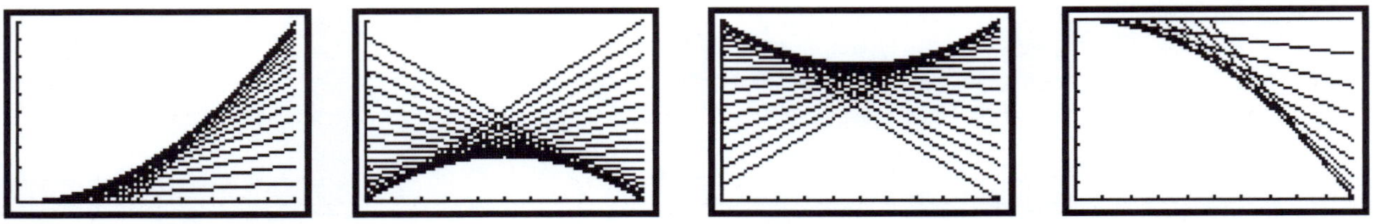

So finally ...

◊ *Here is some of the mathematical jargon that was introduced in this unit:*

δx	$\frac{\delta y}{\delta x}$	$\frac{dy}{dx}$	derivative
limit	local maximum	local minimum	discontinuity

Make sure you understand all these terms before moving on.

Unit 6 nDeriv – how does it do it?

In Unit 5 you saw how to use **dy/dx**, one of the commands in the calculator's CALC menu.

The symbol **dy/dx** was said to stand for the derivative of *y* with respect to *x* and was defined as:

the limit of $\frac{\delta y}{\delta x}$ as δx approaches zero.

In this unit you will use the calculator command **dy/dx** in a different form, **nDeriv**. This form lends itself to more powerful mathematics in later units.

This unit also deals with how the calculator actually works out derivatives – you will be asked to compare gradients calculated by the traditional method developed in Units 3 and 4, with those produced by **dy/dx** and **nDeriv**.

(1) Numerical Derivatives

The calculator's alternative facility that allows you to calculate derivatives or gradients is option 8 in the MATH menu. It is labelled **nDeriv**, standing for numerical derivative.

Clear the home screen and then press [MATH].

Scroll down to find option 8. Notice that it is followed by a bracket to indicate that it requires arguments. Usually three arguments are required:

firstly a function such as **Y1** or **X²+3**;

secondly a variable, usually **X**;

thirdly a value at which the derivative or gradient is to be calculated.

So, to calculate the derivative, with respect to X, of the function stored in Y1 at the point where X=1, you would enter **nDeriv(Y1,X,1)**.

Enter this now. The key presses needed are

[MATH] **8** [VARS] [▸] **1 1** [,] [X,T,Θ,n] [,] **1** [)]

Assuming you still have the cubic function from Unit 5 stored in Y1, when you press [ENTER] you should see the result shown here. This is the same approximation for the gradient of the curve at X=1 as you got using **dy/dx**. (See Unit 5, Activity 2.)

(2) Practice with nDeriv

(a) Use **nDeriv** to calculate estimates for the gradients of the curve **Y1=-.15X³+.1X²+1.6X** at the following values of X.
Give your answers correct to 2 decimal places.

X = 2.0   X = 2.2   X = 2.4   X = 2.6   X = 2.8   X = 3.0

(b) Use **nDeriv** to find the gradient of X^2–X at X = 5, X=50 and X = 500.

(3) Guess before you press with gradients

Use your knowledge of the shape of the graphs of the following functions to guess, as best you can, what values are produced when the following gradient calculations are entered on the calculator. Check your guesses by entering the expressions.

(a) **nDeriv(X², X, 0)**   (b) **nDeriv(X³, X, 0)**

(c) **nDeriv(X² + 5, X, 2)**   (d) **nDeriv(X³ – 3, X, ⁻3)**

(e) **nDeriv(3X – 5, X, 2)**   (f) **nDeriv(4 – X, X, ⁻3)**

(g) **nDeriv(5, X, 5)**   (h) **nDeriv(sin(X), X, 0)**

(i) **nDeriv(X², X, 1)**   (j) **nDeriv(A², A, 1)**

(4) How does it do it? What increment does it use?

The rest of this unit helps you to understand how the calculator produces good approximations to the value of **dy/dx** and **nDeriv**.

Recall that **dy/dx** is defined as the limit of $\delta y/\delta x$ as δx approaches zero. The calculator must use an incremental value very close to zero in its calculation, but it cannot use the actual value zero. **nDeriv** may be used with a fourth optional argument that determines the incremental value (close to zero) of the variable which is used for the calculation. When this fourth argument is omitted, the calculator uses a pre-determined default incremental value.

So can you determine this default increment?

This can be investigated by comparing values for **nDeriv** produced by various fourth arguments, with the value of **nDeriv** when the fourth argument is omitted.

(a) Start by checking that **Y1=-.15X³+.1X²+1.6X** is still entered on the Y= screen. Now move to the list screen and clear L1, L2 and L3 if necessary.

(b) On the home screen enter **nDeriv(Y1,X,1,E⁻1)→L3(1)**. This evaluates the approximate gradient at X=1, using the increment E⁻1 or 0.1. It also stores the result in L3(1). (You will see later why it is useful not to use L1 or L2 here.)

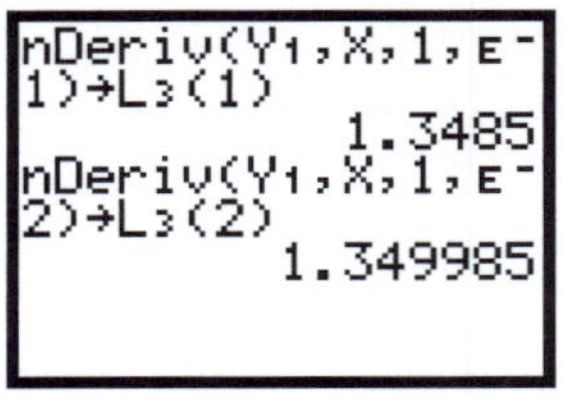

(c) Now enter **nDeriv(Y1,X,1,E⁻2) →L3(2)**. (The easy way to do this is to use [2nd] [ENTRY] and edit the expression.)

Investigate what happens as you decrease the incremental value of the variable by changing to E⁻3, E⁻4, etc., storing the results in successive cells of L3.

(d) Compare your results with the value calculated by **nDeriv(Y1,X,1)** in Activity 1. Can you see what value was used for the default increment?

(e) Investigate what happens if you use larger increments such as 1 or 2.

(5) How does it do it? Perhaps the traditional method?

You have seen that the default increment used by the calculator is E⁻3 (or 0.001).

The next step is to check whether the calculator uses essentially the same traditional method for its calculation of **nDeriv** as was developed in Unit 4.

There you used the formula: $$\text{gradient} = \frac{\text{Y1(X+ }\Delta\text{Tbl) – Y1(X)}}{\Delta\text{Tbl}}.$$

Use program **GRADY** once again to calculate the values of the gradient at X=1. It uses the formula above, with decreasing values of **ΔTbl**. The program stores the values of **ΔTbl** in L1 and of the gradient in L2.

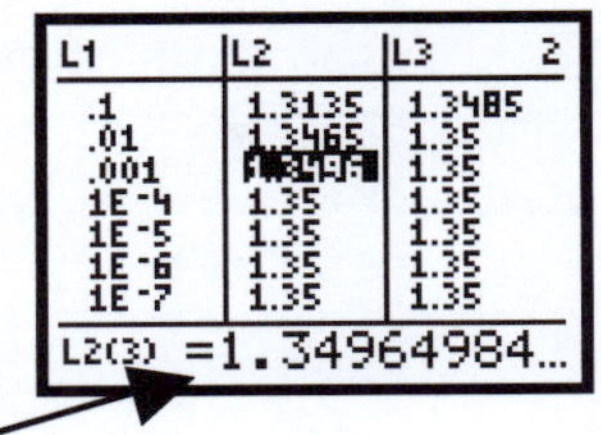

Move to the list screen and you should see the display shown on the right. List L2 holds values of the numerical derivative calculated by the traditional method and L3 holds corresponding values produced using the calculator's **nDeriv** function in Activity 4.

By moving the cursor over any particular list item, its more accurate value is shown at the bottom of the screen. The accurate value of L2(3) is shown here.

Inspecting corresponding values in L2 and L3, it becomes clear that there are differences.

For example, when the calculator uses an incremental value of 0.001 the value of **nDeriv** at X=1 is given in L3(3) as 1.34999985. However, with δx (or **ΔTbl**) = 0.001 the calculator produces a value in L2(3) of 1.34964984 – a small but measurable difference!

So the calculator must use a different method from the traditional one.

(6) How does it do it? Is there an alternative method?

Consider again the traditional method for the calculation of gradient as used in Unit 4. It can be expressed as

$$\text{gradient} = \frac{\text{Y1}(\text{X}+\delta\text{x})-\text{Y1}(\text{X})}{\delta\text{x}}$$

(using δx as an alternative way of writing **ΔTbl**).

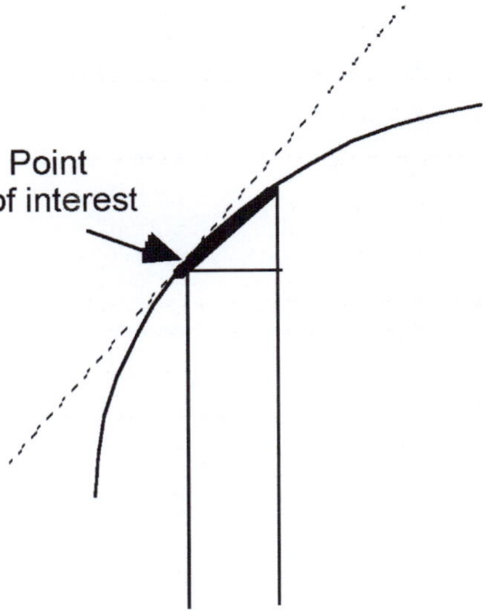

This method uses the gradient of the line joining the point of interest to another point on the curve just above it as shown in the diagram. This line is shown in bold and its gradient is an approximation to the gradient of the tangent *at* the point of interest, which is shown by a dotted line in the diagram.

However, as you saw in Activity 5, this cannot be the method used by the calculator.

The method the calculator uses to find **nDeriv** is known as the "symmetric difference quotient method". Rather than using the traditional formula,

$$\text{gradient} = \frac{Y1(X+\delta x)-Y1(X)}{\delta x},$$

it uses this one:

$$\text{gradient} = \frac{Y1(X+\delta x)-Y1(X-\delta x)}{2\delta x}$$

Compare the two formulas carefully now.

This method uses the gradient of the line joining two points whose X-coordinates are equally spaced either side of the point of interest as shown in this diagram. Although the bold line does not actually pass through the point of interest, its gradient is likely to be a better approximation to the gradient of the dotted line. Compare this with the diagram representing the traditional method on the previous page.

The calculator uses the symmetric difference quotient method because, for most points, it provides a closer approximation to the gradient at the point of interest.

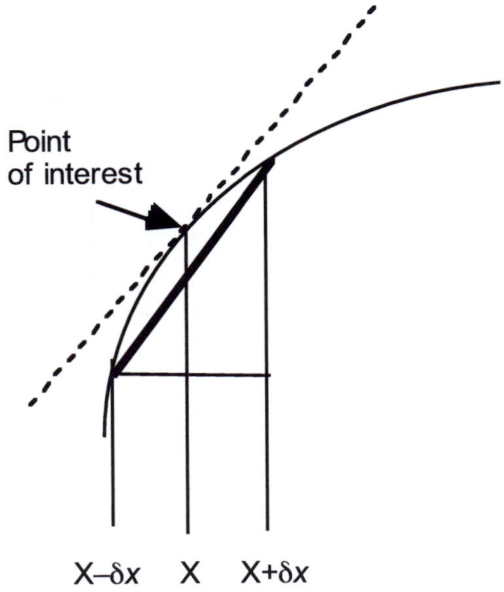

Look closely at the values in lists L2 and L3. Bearing in mind that the true gradient at X=1 is 1.35, can you see that the symmetric difference quotient method (values in list L3) provides a better estimate than the traditional method (values in list L2) for any particular incremental value?

Brain stretcher **Trad v SDQ**

For what shapes of graph does the traditional method for estimating gradients produce an underestimate and for what shapes an overestimate?
Can it ever produce a perfect result?
Can it ever produce an estimate that is totally wrong?

It is rather harder to answer the same questions for the symmetric difference quotient method. Can you do it?

So finally ...

◊ *Why do you think there are two calculator commands for calculating gradients,* ***dy/dx*** *and* ***nDeriv****? In what circumstances is each one useful?*

◊ *In this unit you found out how the calculator has been programmed to calculate gradients. Do you think this has given you any new insights into gradients?*

Unit 7 Gradient functions

In Unit 6, Activity 3, you calculated the gradient of a curve at various points and, as expected, found that the gradient had a range of different values. It is this feature that distinguishes a curve from a straight line – for a curve the gradient is constantly changing, whereas any straight line has a constant gradient. It is possible to think of the gradient of a function as itself forming a new function, known as the *gradient function* or *derived function* and denoted by $\frac{dy}{dx}$.

In this unit the focus is on the gradient functions of cubic functions such as

Y1=-.15X³+.1X²+1.6X.

Is there a way of calculating what the derived function, $\frac{dy}{dx}$, for such a cubic will be? The process of calculating derived functions is known as *differentiation*.

(1) From the particular to the general

Consider once more the function $y = {}^{-}.15x^3 + .1x^2 + 1.6x$.

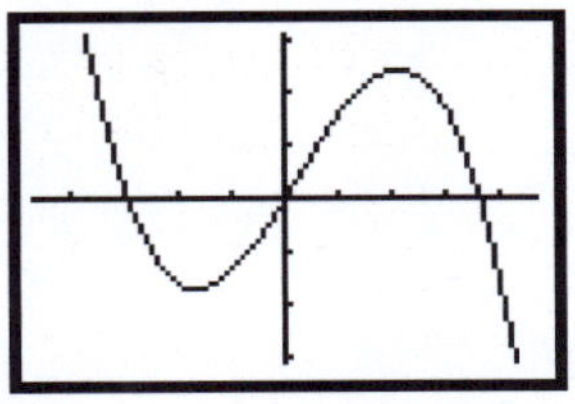

Notice that the highest power of x in this function is 3 (the first term is $^{-}.15x^3$) and it is therefore a member of the family of functions known as *cubics*.

If necessary, re-enter the equation on your calculator as Y1 and draw the graph of the function using the **ZDecimal** window settings.

You saw in Unit 5 that the gradient at any particular point could be calculated on the graphing screen by using option 6 in the CALC menu: **dy/dx**. In Unit 6 you saw that there was an alternative way of calculating the gradient at any particular point, using option 8 in the MATH menu, **nDeriv**, for example by entering **nDeriv(Y1, X, 1)**.

Both of these methods calculate the gradient at one *particular* point. What is required now is to find the gradient for some *general* point.

This is a fundamental mathematical process – *moving from the particular to the general.*

The **nDeriv** command enables this generalisation to take place by replacing the *particular* x-value in the third argument (1 in the example above) by the *general* variable, X, to form **nDeriv(Y1, X, X)**.

Try entering **nDeriv(Y1, X, X)** on the home screen now.

You may not get the value shown here, but the result is certainly NOT a gradient *function*! It is a particular gradient value.

Try to explain what is going on.

(2) The value of X on the home screen

When you enter **nDeriv(Y1, X, X)** on the home screen, the calculator says to itself

> "Ah, I've got to evaluate **nDeriv** of **Y1** with respect to **X** for the *current* value of **X**. Now **X** is 4.7 (or whatever) at the moment so **nDeriv** is....".

Check this by going to the graphing screen and using ◀ or ▶ to change the X value. Then return to the home screen and press ENTER to carry out **nDeriv(Y1, X, X)** again. You should see a different calculated value for the gradient. Repeat this several times.

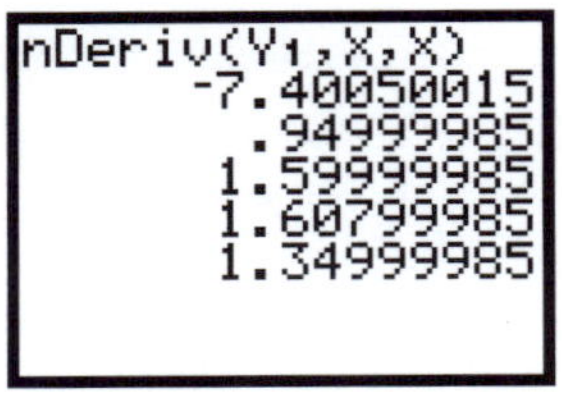

(3) nDeriv on the Y= screen

Now you can make the calculator work much harder and carry out the calculation for *all* the X-coordinates of each pixel on the graphing screen. To do this the **nDeriv** expression needs to be entered on the Y= screen, so forming a new function, the derivative function.

On the Y= screen enter **Y2=nDeriv(Y1, X, X)**.

Press GRAPH and you should see the graphs of Y1 and Y2 displayed together.

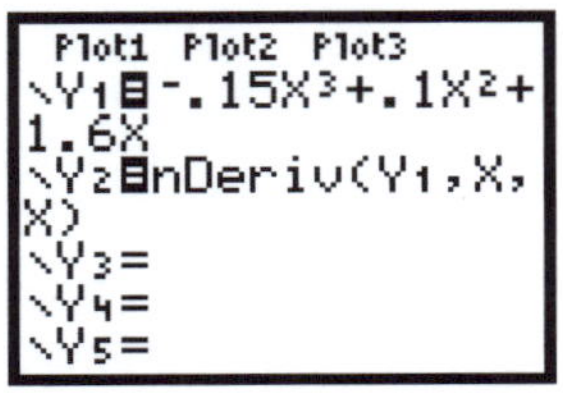

(4) The cubic and its gradient function compared

Look carefully at the graph of the gradient function (Y2) and convince yourself that it is as expected.

- Look at where Y2 cuts the horizontal axis, that is where Y2=0. Is the gradient of Y1 actually zero at that value of X?
- Look at where Y2 is greatest. Does this correspond to the steepest part of the graph of Y1?
- Look at where Y2 is negative, that is below the horizontal axis. For those values of X, is the gradient of Y1 negative (that is, does the graph of Y1 slope downwards)?

(5) What's the equation of the gradient function?

In order to concentrate solely on the gradient function for a while, return to the Y= screen and deselect Y1.

Press GRAPH again and think about the shape of the graph.

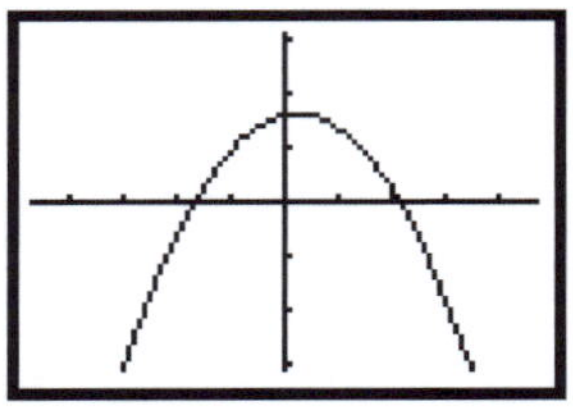

Write a short description of this graph, mentioning its most important features. If you can, make a stab at saying what its equation might be.

(6) A strategy for finding the equations of graphs

You have seen that the use of **nDeriv(Y1,X,X)** enables you to draw the graphs of a derivative function. However, often it is necessary to know the equation of such a function. Some more advanced calculators can do the algebraic manipulation necessary to produce the equations of derived functions; that is to differentiate. However, although the members of theTI-83 family of calculators do not carry out such differentiation directly, there is a method of fitting an equation to a graph and that procedure is described in the next three activities.

There is a range of powerful commands built into the calculator. Known as the regression facilities, they form a large part of the STAT CALC menu.

Press [STAT] [▶] and scroll down to see the list of these facilities.

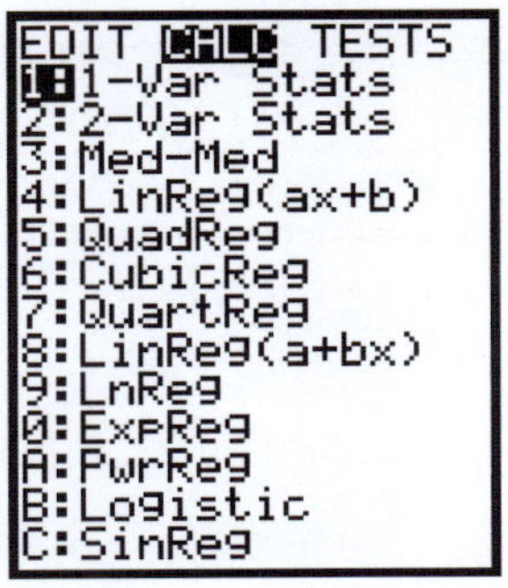

In **Calculator Maths: Statistics** some of these facilities are explained in more detail, but that is not repeated here. All you need to know at the moment is what the commands will do and how to use them.

Essentially the commands make the calculator fit the best possible line or curve to a scatterplot. So, if you have a scatterplot with points that look as if they lie on a straight line, option 4, **LinReg(ax+b)** would calculate the equation of the line that would fit the points best.

The curve that you produced in Activity 5 looks very much like the graph of a quadratic equation so option 5, **QuadReg** would be a good possibility. However, this needs a scatterplot to work on and at the moment all you have is a function defined by **Y2=nDeriv(Y1, X, X)**. Can a scatterplot be produced from this function?

Fortunately, this is possible using a similar technique to that used in Unit 2, Activity 3. (You might find it useful to glance back at that activity now.) The idea is to store in one list a sequence of X-coordinates from the current graphing screen, then use the function in Y2 to calculate the corresponding Y-coordinates and store them in a second list. Finally simply plot the data in the two lists as a scatterplot.

So to summarise, the strategy to find the equation of the gradient curve produced by the function in Y2 involves the four steps shown here.

▶▶ Step 1 Enter a sequence of suitable X values in L1, using the **seq** command.

▶▶ Step 2: Enter the corresponding Y values in L2, using **Y2(L1)**.

▶▶ Step 3: Produce a scatterplot of the values in L1 and L2.

▶▶ Step 4: Use **Quadreg** to find the equation of the curve.

(7) First three steps: a scatterplot from the function

Because this is a technique that will be used frequently later in the book, it is worth creating a program at this stage.

Look carefully at the program **SCATTER** shown here and compare it with what you did in Unit 2, Activity 3.

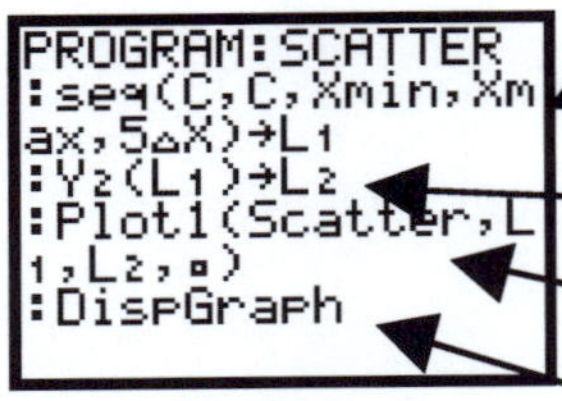

Step 1. Store the X-coordinate of every 5th pixel in L1.

Step 2. Store corresponding Y-coordinates in L2.

Step 3. Set up a scatterplot and display it.

Create the program **SCATTER** now – the key presses are below if you need them.

[PRGM] [◄] **1 SCATTER** [ENTER]

[2nd] [LIST] [►] **5** [ALPHA] **C** [,] [ALPHA] **C** [,] [VARS] **1 1** [,] [VARS] **1 2** [,] **5** [VARS] **1 8** [)] [STO►] [2nd] [L1] [ENTER]

[VARS] [►] **1 2** [(] [2nd] [L1] [)] [STO►] [2nd] [L2] [ENTER]

[2nd] [STAT PLOT] **1** [2nd] [STAT PLOT] [►] **1** [,] [2nd] [L1] [,] [2nd] [L2] [,] [2nd] [STAT PLOT] [►] [►] **1** [)] [ENTER]

[PRGM] [►] **4**

Press [2nd] [QUIT] to leave the program editor and execute the program **SCATTER**.

You should see points plotted all along the graph of Y2. (The coordinates of these points are stored in L1 and L2.)

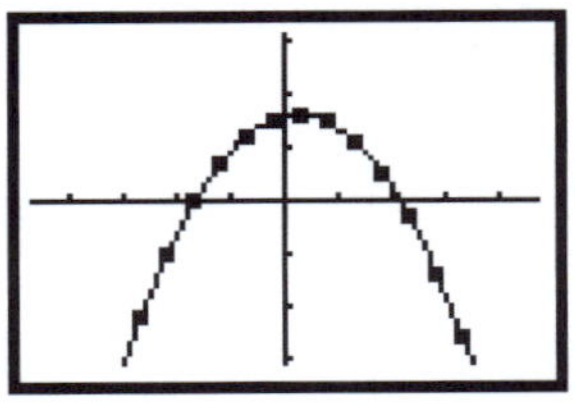

(8) Last step: use QuadReg

The next stage is to carry out a quadratic regression calculation on the data in L1 and L2.

To do this, go to the home screen and enter **QuadReg L1, L2,Y3** by pressing:

[STAT] [►] **5** [2nd] [L1] [,] [2nd] [L2] [,] [VARS] [►] **1 3** [ENTER]

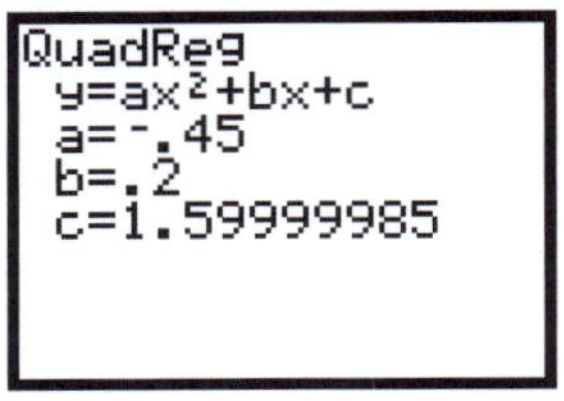

The resulting screen shows that the quadratic equation which fits the data in L1 and L2 best is $y = {}^{-}.45x^2 + .2x + 1.6$ (rounding the value of **c** to one decimal place.)

Press [Y=] and confirm that this equation has been stored as Y3.

(9) Checking the quadratic equation

Is the quadratic function that you have calculated and stored in Y3 equivalent to the gradient function that is stored in Y2?

(a) Investigate the graphs of the two functions by tracing along them. Are there any significant discrepancies?

(b) Investigate the tabulated values of the two functions using [2nd] [TABLE]. Are there any significant discrepancies?

(10) Time to recap and to ponder

So what does all this show? Here is a summary of the story so far.

- You entered in Y1 the cubic function

  $$y = {}^-.15x^3 + .1x^2 + 1.6x.$$

- You entered its derivative or gradient function, $\frac{dy}{dx}$, in Y2 and this appeared to be a quadratic function.
- You have shown that this derivative function is actually the quadratic function

  $$\frac{dy}{dx} = {}^-.45x^2 + .2x + 1.6$$

Is there a connection between the coefficients in the two functions in Y1 and Y3?
Is it pure coincidence that 1.6 appears in both?
What about the other coefficients?

Write down any thoughts you have about the coefficients in these two functions.

(11) Time to investigate

A mathematical investigation seems to suggest itself. With very little effort it is possible to change the coefficients in Y1 and see the effect on the derived function by repeatedly using program **SCATTER** and quadratic regression.

You may like to try this investigation for yourself now.

Be very systematic in the way you change the coefficients in Y1 and see whether you can come up with a general rule to calculate the derived function of *any* cubic expression.

However, if you feel that you need some guidance, work your way through the following activity.

(12) Cubic differentiation

a) On the Y= screen, clear Y3.

In function Y1, change the coefficient of X from 1.6 to some other number.

Return to the home screen and execute the program **SCATTER**. (You can use [2nd] [ENTRY] to avoid having to enter this again.) You should see the graphs of the new function (in Y1) and its new derived function (in Y2), together with the points produced by the scatterplot.

Now return to the home screen and use [2nd] [ENTRY] to enter **QuadReg L1, L2, Y3**.

What do you notice about the value of **c** that is produced by the quadratic regression calculation?

What if the 1.6 in the original Y1 were changed to ⁻1.6 or some other negative number?

(b) Now try changing the coefficient of X^2 in the original function from .1 to some other positive or negative number. Repeat the process of using **SCATTER** and quadratic regression.

What do you notice about the value of **b**? Try expressing the rule very precisely in words.

(c) Now try changing the coefficient of X^3 from ⁻.15. Once you have noticed what happens to the value of **a**, express the rule precisely in words. Can you express the rule using symbols?

(d) What would be the effect of adding an extra constant term onto the end of the expression in Y1? For example, what effect would there be on the derived function if the original function in Y1 were

$$y = {}^{-}.15x^3 + .1x^2 + 1.6x + 8.$$

Can you explain why this should be?

(e) Put together the results from parts (a)-(d) to answer the following question.

If the original cubic function were $y = px^3 + qx^2 + rx + s$ what would be the equation of the derived function?

So finally ...

This unit has taken some big steps forward. Spend some time now looking back and taking stock.

◊ *Previously you were looking at gradients at particular points, but now you have met functions that are themselves the gradients of other functions. (Activities 1 to 3)*

◊ *You have also seen a strategy by which your calculator can be used to work out the equation of some gradient functions. (Activities 4 to 9)*

◊ *You have formulated a general rule for differentiating cubic functions. (Activities 10 to 12)*

Unit 8 Differentiation of polynomials

In the last activity of Unit 7 you should have found that for any cubic function of the form:

$$y = px^3 + qx^2 + rx + s,$$

$$\frac{dy}{dx} = 3px^2 + 2qx + r$$

where p, q, r and s are constants.

This cubic function is a particular type of *polynomial* – an expression of two or more terms containing different powers of the same variable.

In this unit you will see how to differentiate functions such as $y = x^6$, leading to a general rule for gradients of polynomials.

(1) Differentiating a quadratic

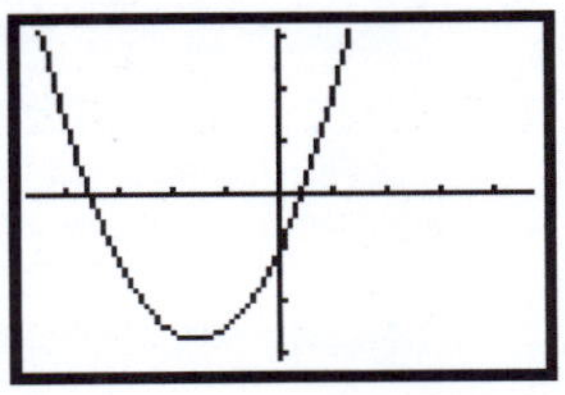

Determine the derived function of $y = 0.7x^2 + 2.2x - 1$ using the following three methods.

(a) Even though this is not a cubic function it can be thought of as one with $p = 0$.

Use the result stated above to write down the value of $\frac{dy}{dx}$.

(b) Enter the function on the Y= screen as Y1. If necessary, enter the function **Y2=nDeriv(Y1, X, X)**, and deselect it. Clear any other functions on the Y= screen. Switch off statistical plots using option 4 in the STAT PLOTS menu, **PlotsOff**.

Using the **ZDecimal** window settings, display the graph of the function Y1 and check that, as expected, it has the distinctive shape of a quadratic.

Execute the program **SCATTER**, and use **QuadReg** to calculate the derived function, storing the result in Y3. This should confirm your answer to part (a), though notice that there is an X^2 term with a coefficient of zero.

(c) The derived function is actually a straight line. This means that instead of using *quadratic* regression to find the line of best fit for the scatterplot it would have been more obvious to use *linear* regression.

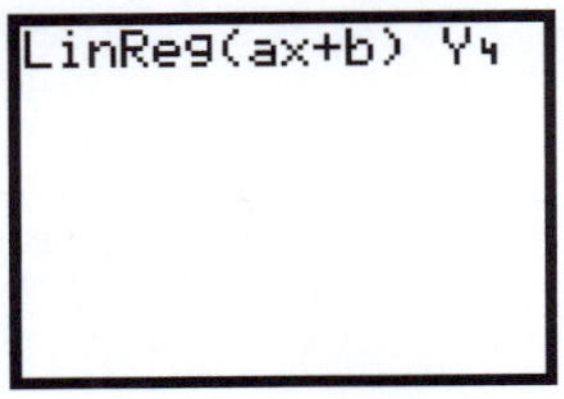

Carry out a linear regression calculation and store the result as Y4, by entering **LinReg(ax+b) Y4**. Check that this produces the same result as you had in parts (a) and (b).

(2) Differentiating a linear function

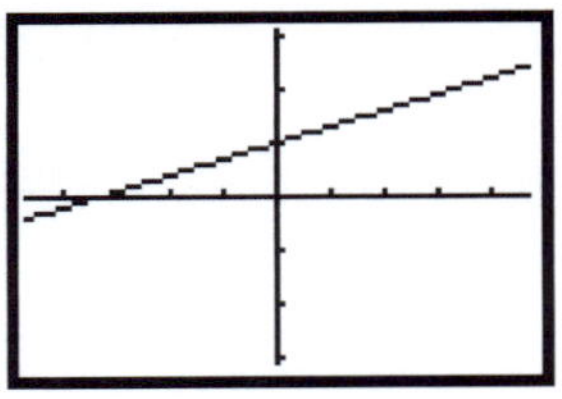

Determine the derived function of $y = 0.3x + 1$ using the following methods.

(a) Although it is a linear function, $y = 0.3x + 1$ can be thought of as a cubic function with both $p = 0$ and $q = 0$.

Use the result stated on page 39 to write down the value of $\frac{dy}{dx}$.

(b) Enter the function as Y1 and use the program **SCATTER**. How does this confirm your answer to part (a)?

(c) Use an appropriate regression calculation to fit a function to the points displayed on the scatterplot.

(3) Pattern spotting

Look again at the rule developed in the previous unit for differentiating a cubic:

If $y = px^3 + qx^2 + rx + s$,

$$\frac{dy}{dx} = 3px^2 + 2qx + r$$

Can you guess the rule for differentiating a quartic – that is a function where the highest power of x is 4?

(a) Write down your guess for $\frac{dy}{dx}$ if

$$y = ax^4 + px^3 + qx^2 + rx + s$$

(b) Use your guess to predict $\frac{dy}{dx}$ for the particular case

$y = 0.1x^4$ (i.e. p, q, r and s are all zero.)

(c) On your calculator enter **Y1 = 0.1X^4** and switch on the function **Y2=nDeriv(Y1, X, X)**. Does the graph of the derived function support the prediction you made in part (b)?

(d) Enter as Y3 your prediction for $\frac{dy}{dx}$ and see if it has the same graph as Y2.

(e) Finally, for further confirmation, use the program **SCATTER** and **CubicReg** (option 6 of the STAT CALC menu) to calculate the equation of the derived function.

(4) The gradient of $y=x^5$

Option 7 of the STAT CALC menu is **QuartReg** (standing for Quartic Regression).

Use this to calculate the equation of the gradient function of $y=x^5$.

(5) A general rule

By now you should be able to complete the table below (*a* is a constant).

$y = ax$	$y = ax^2$	$y = ax^3$	$y = ax^4$	$y = ax^5$	$y = ax^6$
$\frac{dy}{dx} =$	$\frac{dy}{dx} =$	$\frac{dy}{dx} =$	$\frac{dy}{dx} =$	$\frac{dy}{dx} =$	$\frac{dy}{dx} =$

So you can now move to a general rule: if $y = ax^n$ then $\frac{dy}{dx} =$?

However, just because the rule is true if n = 1, 2, 3, 4 and 5, it doesn't mean that it is necessarily true for all values of *n*.

Test the rule for each of the following values of *n* by entering the basic function in Y1, and comparing the derived function predicted by the rule (entered in Y3) with the function **Y2=nDeriv(Y1, X, X)**.

(a) $n = 8$ (b) $n = 20$ (c) $n = 0$ (d) $n = {}^-2$

(6) More tests for the rule

Here is a slightly different way of testing the general rule.

Produce these screen displays on your calculator.

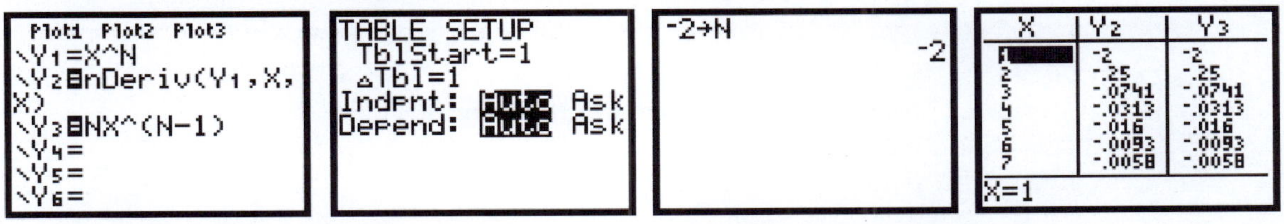

So far you have tested the general rule when *n* has various positive and negative integers.

Now test it for the following values of of *n*.

(a) $n = 0.625$ (b) $n = \frac{1}{3}$ (c) $n = \sqrt{2}$ (d) $n = {}^-2.5$

So finally ...

So now you have a general rule for differentiating functions that are powers of x:

when $y = ax^n$ for any value of n and a is a constant, $\frac{dy}{dx} = nax^{n-1}$.

And in previous units you have seen examples illustrating that, when you have a polynomial consisting of the sum of several terms, the derived function is the sum of the individual derived terms.

So, for example, if $y = 5x^4 - 2.1x^2 + 6$, then $\frac{dy}{dx} = 20x^3 - 4.2x$.

But....are you convinced?

Unit 9 Points of inflexion

In previous units you have seen that differentiating a function enables you to determine exactly the function's local maximum and minimum points – they occur where the derived or gradient function is zero.

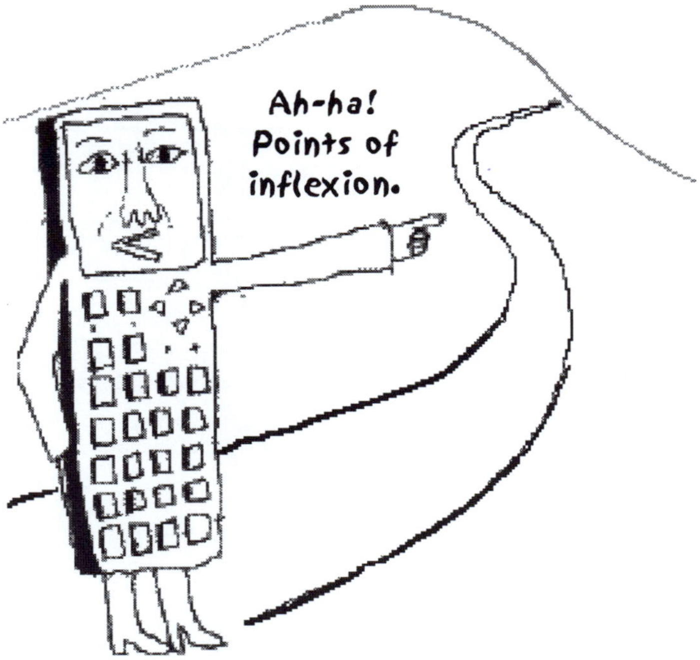

Many graphs have other important points called points of inflexion. For example, there is a point of inflexion where the road shown here stops curving to the left and starts curving right.

Similarly, imagine walking up the profile of this hill (from right to left). The climb gets steeper and steeper until, at about where the road disappears over the hillside, the gradient begins to reduce again – another point of inflexion.

In this unit you will find out how to use your calculator to calculate the coordinates of points of inflexion. Whereas the calculator has options in the CALC menu that allow you find maximum and minimum points, there is no similar pre-programmed command for points of inflexion – yet!

(1) The gradient of the gradient

(a) For one last time on the Y= screen enter, if necessary, the cubic equation **Y1=$^{-}.15X^3+.1X^2+1.6X$**.

Also enter **Y2=nDeriv(Y1, X, X)**.

With the **ZDecimal** window settings, press [GRAPH] to see the cubic function and its gradient function displayed together.

(b) Now think about the gradient-of-the-gradient function.

How does the gradient of Y2 vary?
What would you expect its gradient function to look like?
Draw it in your mind's eye.

Now enter **Y3=nDeriv(Y2, X, X)** and see if your mind picture was right. (You do have to be a bit patient – just think how many calculations the machine is having to do in order to plot every pixel-point!)

Just as the gradient function is called the derivative and is written $\frac{dy}{dx}$, so the gradient-of-the-gradient function is called the *second derivative* and is written $\frac{d^2y}{dx^2}$.

(2) Comparing the curves

(a) Now deselect Y1 and compare the graph of Y2 (the gradient function) with Y3 (the gradient-of-the-gradient function). As you would expect, where Y2 has a local maximum, its gradient is zero so Y3 crosses the x-axis at the corresponding X value. Trace along Y3 and find the approximate value of X at which this occurs.

Also use the Table command to find (correct to 3 d.p.) the value of X at which Y3 is zero.

(b) Use **maximum** in the CALC menu to determine where Y2 has a local maximum.

(c) Use **zero** in the CALC menu to determine where Y3 cuts the X-axis, to confirm your answers in parts (a) and (b).

(d) Finally re-select Y1 and trace along the cubic graph close to the value of X that you found in the earlier parts of this question. It should be clear that this point is certainly not a local maximum or minimum, nor does it lie on either of the axes. However, it is still a significant point – a point of inflexion.

(3) Reflection on inflexion

Look carefully at the point of inflexion on the graph of Y1 and think about each of these alternative ways of describing it.

(a) The point of inflection is where the graph is locally steepest. Above and below this point it is less steep.

(b) Imagine steering a car along a road represented by the curve. At the point of inflexion the steering wheel will, for an instant, be pointing straight ahead.

(c) A point of inflexion is where the curve changes from being concave to convex (or from convex to concave).

(d) A point of inflexion is where the second derivative is zero.

(4) Second derivative – without a calculator!

(a) In Unit 7 you saw that if $y = {}^{-}.15x^3 + .1x^2 + 1.6x$

the gradient or derivative function, $\frac{dy}{dx} = {}^{-}.45x^2 + .2x + 1.6$.

Now differentiate *this* function, using the rule for differentiating polynomials, to find the gradient-of-the-gradient function, or second derivative function, $\frac{d^2y}{dx^2}$.

(b) Put $\frac{d^2y}{dx^2} = 0$, and solve the equation to find the values of X at the point of inflexion.

(5) A different cubic

Enter **Y1= X³/12+X²/3–X/4–1.5**.

Use one of the methods in Activity 2 to find the point of inflexion.

Now use the method of Activity 4 to find the point of inflexion.

(6) A quartic

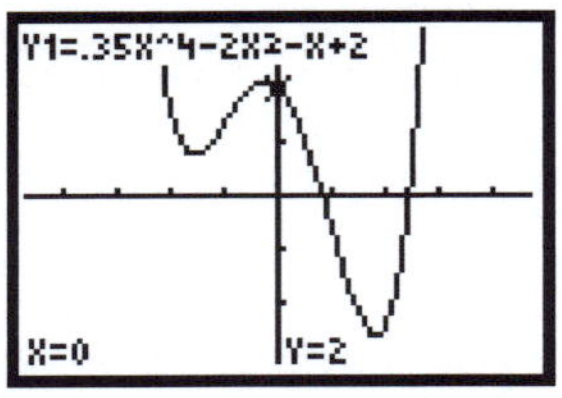

Look at the interesting curve shown on the right. How many points of inflexion can you see?

What do you think the graphs of its derived function and its second derived function will look like?

Enter the function in Y1 and use a calculator method to find the points of inflexion.

Now use the method of Activity 4 to find the points of inflexion.

(7) A program for points of inflexion

Unlike local maxima and minima, there is no direct way to find a point of inflexion on your calculator, but you can remedy this by writing a short program.

As with any program, it is necessary to think out what procedure (or algorithm) is necessary before starting to enter the commands. Below is a summary of the steps that were used earlier in this unit to find points of inflexion of functions stored in Y1. Alongside are comments about how each step can be programmed.

Step 1 On the Y= screen, enter **nDeriv(Y1,X,X)** into Y2 and **nDeriv(Y2,X,X)** into Y3.	The programming commands need to have the functions stored inside quotation marks: e.g. "**nDeriv(Y1,X,X)**"**->L2**. To keep the graphing screen simple, **FnOff 2,3** can be used to prevent Y2 and Y3 being drawn.
Step 2 Decide which point of inflexion is to be found, since there is sometimes more than one.	In the program this can be done by inviting the user to make a guess where the point of inflexion lies – this will work in a similar manner to the **Guess** feature in some of the options in the CALC menu. (Simply move the cursor to the required position and press [ENTER]). The guess can be stored in G.
Step 3 Find the value of X where Y3 (the second derivative) is zero.	Option 0 in the MATH menu works differently in programming mode: **solve(Y3,X,G)** will find the required value of X.

Here is one version of the program. Create it on your calculator now. The solution at the back of the book gives the required key presses if you need them.

PROGRAM: INFLEX

:"nDeriv(Y1,X,X)" → Y2	Store the derivative of Y1 in Y2.
:"nDeriv(Y2,X,X)" →Y3	Store the derivative of Y2 in Y3.
:FnOff 2,3	Deselect Y2 and Y3.
:Text(6,0,"GUESS?")	Display **GUESS?** near top of graphing screen.
:Trace	Allow the user to select a point on the curve.
:X→G	Store the x-coordinate of the guessed point in G.
:Text(0,0,"POINT OF INFLEXION")	Display **POINT OF INFLEXION** at top of screen.
:Text(6,0,"AT X=", solve(Y3,X,G))	Display **AT X=** followed by the solution of Y3=0.

Press [2nd] [QUIT] to leave the program and use it to check your answers to Activity 6.

(8) Some further points

Use the program to find the points of inflexion of the polynomial functions shown here.

(a)

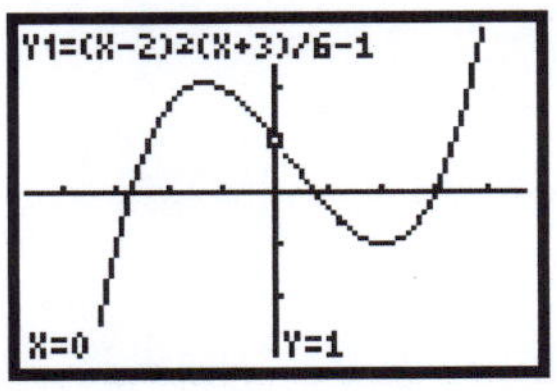

(b)

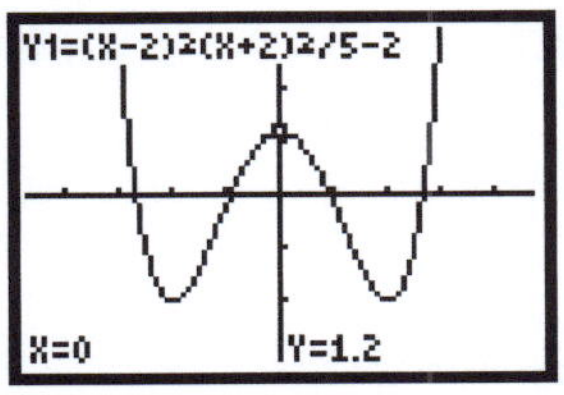

(c)

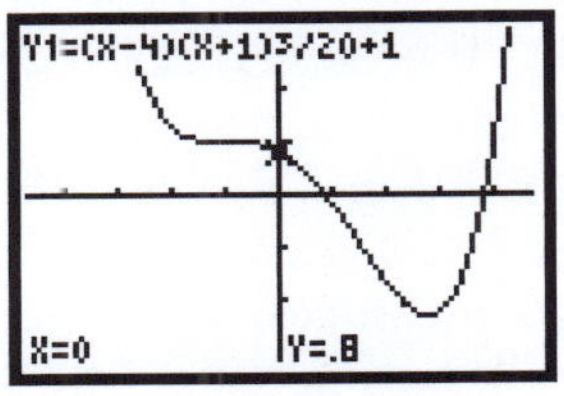

(d)

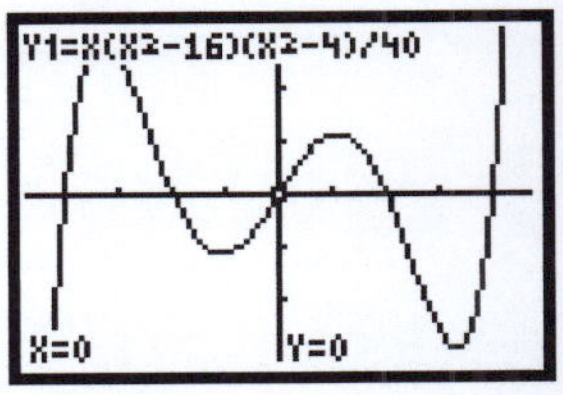

Brain stretcher — **Fixing a bug**

Try entering **Y1=X+2** and running the program **INFLEX**. Is there a problem? And what about **$Y1=X^2–2$**?
Can you explain what is going on by referring to the graphs of Y1, Y2 and Y3.
Can you think of a way of correcting the program to fix the bug?

So finally ...

You should now be able to identify points of inflexion and be able to calculate their x-coordinates either with or without your calculator.

Unit 10 Rates of change

This unit looks at two applications of differential calculus. The first, marginal analysis, is important in the study of economics. The second part of the unit deals with velocity and acceleration, very important ideas in applied mathematics and physics.

In both cases you will meet the phrase "rate of change". The rate at which a variable is changing is closely related to the gradient of the associated graph. When a graph shows values of *y* plotted against *x*, the gradient of the graph, *dy*/*dx*, can be thought of as the rate at which *y* changes with respect to *x*.

Frequently, variables change over *time*. For example, a graph may be drawn showing how the population of a country has changed over a number of years, so *y* might represent the population in millions and *x* might represent time in years. Then *dy*/*dx* represents the rate of change of the population with respect to time, given as a number of millions per year. Velocity and acceleration are both rates of change with respect to time.

However, rates of change are not always defined with respect to time and sometimes you will meet graphs where *x* represents other variables. For example, a graph could represent the temperature in $^{\circ}$C for various heights in km above the earth's surface. Then the gradient of the graph is the rate of change of temperature with respect to height, measured in C per kilometre.

(1) Marginal analysis

In economics the term *marginal* refers to a rate of change; that is, to a derivative. For example, a manufacturing company may need to know marginal costs: that is the approximate cost of producing one more item after a certain number of items have been already produced. The terms marginal revenue and marginal profit can be defined in a similar way.

In Activities 1 to 3 a mythical company, A+B and co., is analysing its production of units of some commodity and X is used to stand for the number of units produced. So each unit may be a single computer system, or a thousand boxes of chocolate, or a million litres of oil, or whatever. In the same way, Y is used to indicate units of currency, perhaps thousands of euros, hundreds of dollars, etc.

```
WINDOW
 Xmin=0
 Xmax=30
 Xscl=1
 Ymin=-50
 Ymax=500
 Yscl=100
 Xres=1
```

(a) The company has produced an equation to represent the cost, C, of producing X units. This is known as a *model* and is the best available approximation.

The cost function is $C=30+7X+0.01X^3$ for $0 \le X \le 30$. This shows there are fixed costs (30), and variable costs ($7X+0.01X^3$).

Draw a graph representing the cost function using the window settings shown here and use the Table to find the cost of producing 5, 10, 15, 20, 25 and 30 units.

(b) Now calculate the corresponding marginal costs.

(2) Marginal costs and mean costs

You saw in Activity 1 that the marginal costs increased as X increased. Recall that marginal costs are the cost of producing one more item at a particular point i.e. the gradient of the graph of $C=30+7X+0.1X^3$.

Another quantity used to analyse production is the mean (or average) cost. For example, the total cost of producing 5 units is 66.25 currency units so the mean cost per unit is 66.25÷5=13.25, whereas the mean cost of producing 10 units is 110÷10=11 currency units and, in general, mean costs are C÷X. So it seems that from X=5 to X=10 marginal costs increase but mean costs fall. Is the same true for all values of X?

(a) If you have not already done so enter **Y2=nDeriv(Y1,X,X)** to calculate the marginal costs and enter **Y3=Y1/X** to calculate the mean costs. Change the window settings so that **Ymin=$^{-}$5** and **Ymax=50** and compare the graphs of Y2 and Y3.

(b) It is a well known principle in economic theory that, at the level of production where the mean cost is a minimum, the mean cost equals the marginal cost. Verify that this is true in this particular case by using options in the CALC menu to calculate the minimum of the mean costs (Y3) and the intersection of Y2 with Y3.

(3) Revenue, profit and marginal profit

A+B and co. have produced a model to represent the revenue, R, generated by the production of X units. Each unit is sold for 30 currency units but there is a discount for large purchases so the revenue function is $R=30X-0.5X^2$ for $0 \le X \le 30$.

(a) Enter the appropriate function as Y4, deselect Y2 and Y3 and, using the window settings of Activity 1, compare the graphs of the cost function and the revenue function.

(b) The maximum profit occurs when the revenue exceeds the costs by the largest amount. Estimate where this might be using the graphs of Y1 and Y4. Now enter **Y5=Y4–Y1** to represent the profit function. Draw the graph of Y5 and calculate its maximum value and at what level of production it occurs.

(c) Finally calculate the actual and marginal profit when 5, 10, 15, 20, 25 and 30 units are produced.

(4) Velocity and acceleration

The position (in relation to a fixed point) of an object moving in a straight line is known as its displacement from that point. It is not the same as distance because displacement can be positive or negative.

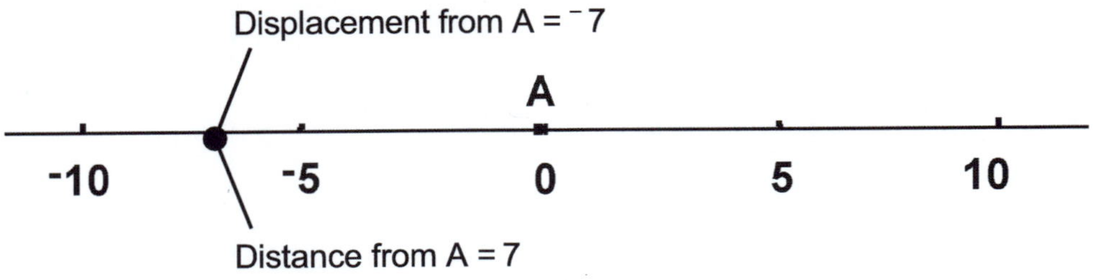

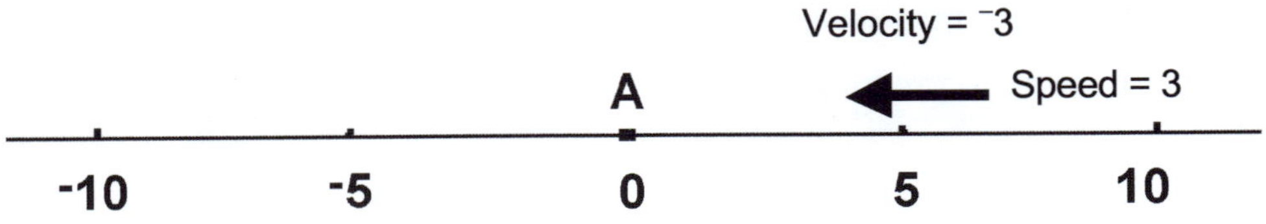

Velocity is defined as the rate of change of displacement with respect to time. It is not the same as speed because velocity can be positive or negative.

So if an object's displacement S is given in terms of time T then its velocity V is given by $V = \frac{dS}{dT}$.

Similarly acceleration (which can also be positive or negative) is given by $A = \frac{dV}{dT}$ or $\frac{d^2S}{dT^2}$.

The following activities give you practice at using your calculator to solve problems involving velocity and acceleration.

(5) I shot an arrow in the air

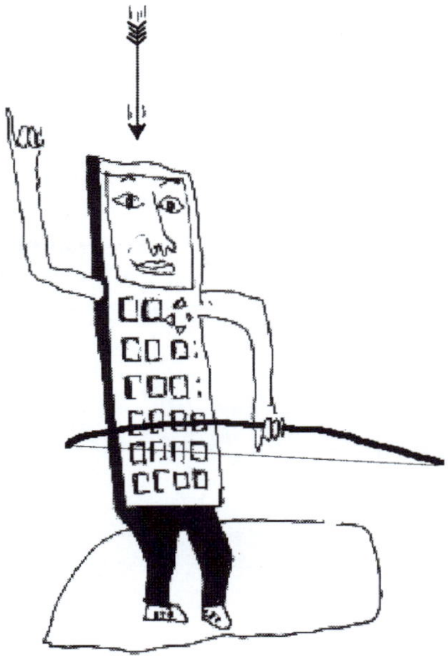

An arrow is fired vertically upwards by a foolish archer and, T seconds after leaving the bow, its height S in metres is given by the equation $S = 22T - 4.9T^2$.

(a) Enter an appropriate equation in your calculator and draw a displacement-time graph to represent the flight of the arrow. Why does the graph *not* show the flight of the arrow as seen by someone watching nearby?

Use options in the CALC menu to answer the questions on the next page.

(b) What is the displacement of the arrow after 2 and 4 seconds?

(c) What is the velocity of the arrow as it leaves the bow string?

(d) After how many seconds does the arrow return to the archer?

(e) What is the velocity of the arrow as it returns to the archer?

(f) What is the maximum height of the arrow?

(g) What is the velocity of the arrow at its highest point?

(h) Produce a velocity-time graph by entering on the Y= screen an equation using **nDeriv**. Explain the shape of this graph.

(i) Calculate the acceleration using *dy*/*dx* at any point on the velocity-time graph.

Brain stretcher **A diversion into ballistics**

Using parametric graphing and one of the hidden options on the Y= screen, it is possible to create a moving image on your calculator's screen to represent the true motion of the arrow in Activity 5. (Though it's more like a cannon ball than an arrow!)

No explanations are given; just follow the visual clues below. After all, it *is* a Brain stretcher.

NB

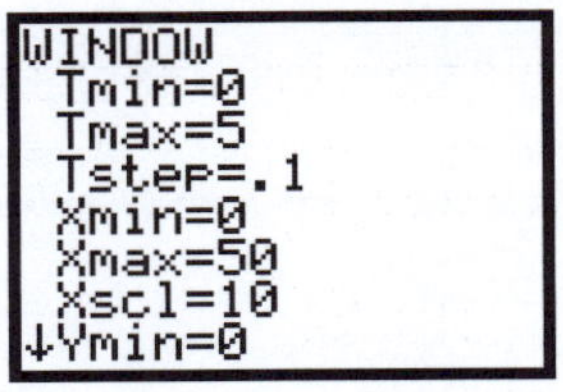

[GRAPH]

[2nd] [DRAW] [ENTER]

NB

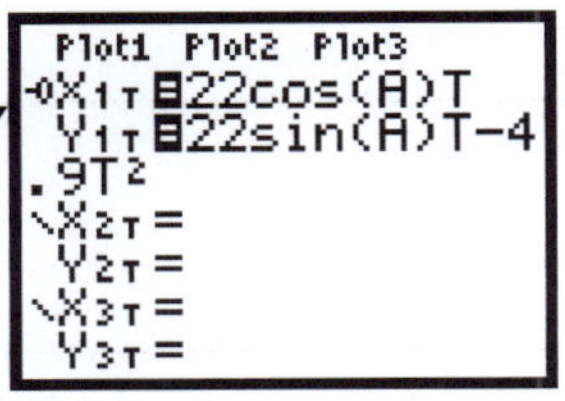

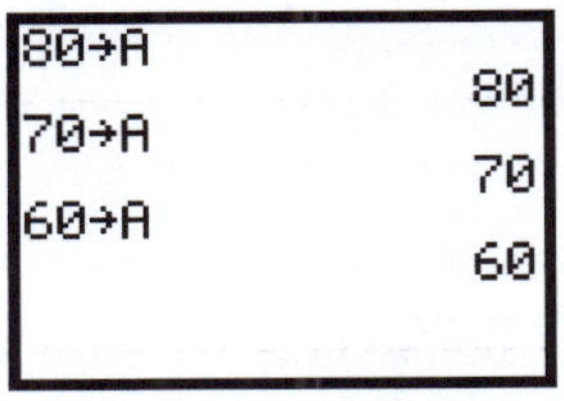

If you have done the Brain stretcher, you should return the MODE menu to its default settings before going on.

(6) Simple harmonic motion

(a) A particle is moving in a straight line and its displacement s from a fixed point is given by the equation $s = a \sin(\omega t + \phi)$. Enter this equation on your calculator in the form **Y1=Asin(WX+F)**.

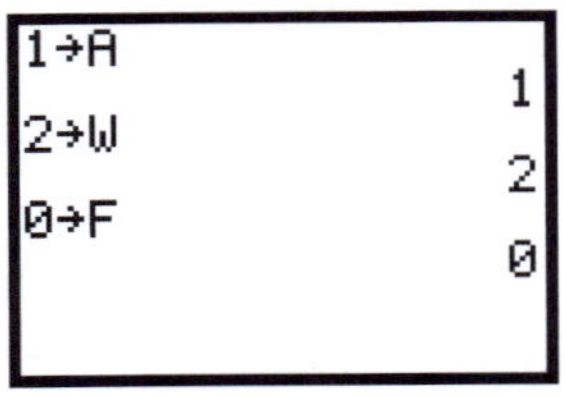

Now store the simple values for *A, W* and *F* shown here.

Make sure that **Radian** is selected on the MODE menu and then use the **ZTrig** window settings to see the displacement-time graph for this motion.

(b) Investigate the effect on the graph of changing the values of *A, W* and *F*. (Hint: try using values of *F* such as π/2 or π.)

(c) Return to the simple values of *A, W* and *F* shown above. Enter the first and second differential functions as shown here. Since Y1 is the displacement function, what functions are displayed by Y2 and Y3?

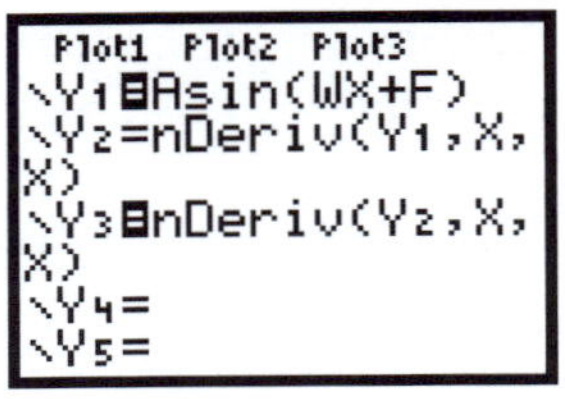

(d) With Y2 de-selected draw the graphs of the displacement and acceleration functions.

(e) A sine curve is a representation of an oscillating pattern known as *simple harmonic motion*. An important feature of simple harmonic motion of the form $s = a \sin(\omega t + \phi)$ is that

acceleration, $\dfrac{d^2s}{dt^2} = {}^-\omega^2 s$.

How do the two graphs that you have just drawn illustrate this relationship?

Try using other values of *W* to see if the relationship holds for them too.

So finally ...

◊ *In this unit you have dipped a toe into Economics and Applied Mathematics to explore two important applications of differentiation. Both used the underlying concept of rates of change.*

In general:

if y is some function of x,
the rate at which y changes as x changes
is the differential of y with respect to x, dy/dx.

◊ *This general principle becomes expressed as follows for the examples you met in this unit:*

if profit, P, is a function of the number of units produced, x,
the rate at which P changes as x changes
is the marginal profit, dP/dx,

and:

if displacement s is a function of time t,
the rate at which s changes as t changes
is the velocity, ds/dt.

Unit 11 Antiderivatives

From this unit on, ***Calculator Calculus*** changes direction. In the first ten units the emphasis has been on one branch of calculus – differentiation. Recall how the early units explored the gradient of functions, first at a particular point and then in general. You saw how to calculate the derivative of functions and used the notation

$\frac{dy}{dx}$ to represent the gradient function.

For many functions of the form $y = f(x)$ you should now be able to calculate $\frac{dy}{dx}$.

Now this book begins to look at a second branch of calculus known as integration. This unit begins by considering what are known as antiderivatives; in other words, if you have $\frac{dy}{dx} = f(x)$, can you work out what function y must be?

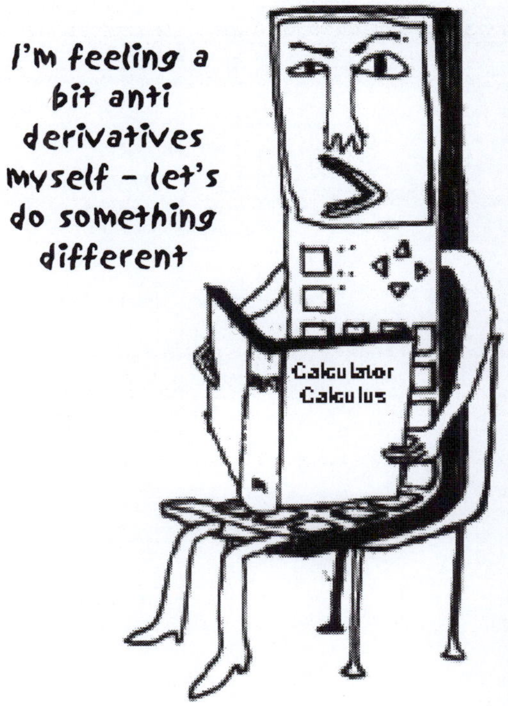

(1) Working back from the gradient

Imagine that your calculator has been set up with some unknown function entered in Y1, but de-selected, and with **nDeriv(Y1,X,X)** selected in Y2. When you press GRAPH, only the gradient function's graph is displayed.

Would you be able to work back from the graph of the gradient function to determine what the graph of Y1 would look like?

Here's a worked example to follow through and then try some others on your own in Activity 2.

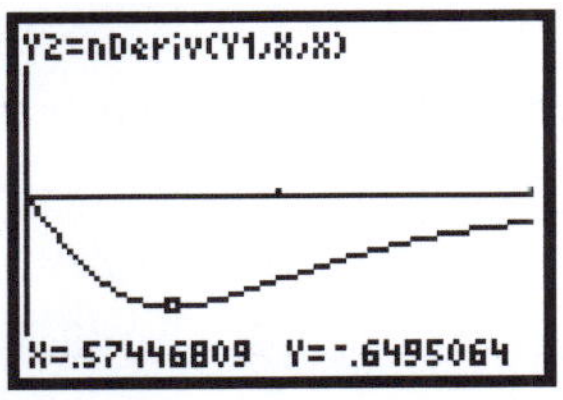

Example: A gradient function's graph is shown on the left.

Since it passes through the origin, the gradient function is zero at X=0. So at this point the original graph must be horizontal.

Between X=0 and X = 0.6 approximately the gradient function is negative and decreasing, so the original graph must slope downwards and is gradually getting steeper.

Between X= about 0.6 and X=2 the gradient function is still negative but increasing, so the original graph must slope downwards and is gradually getting less steep.

So it is possible to make a guess at what the original graph must have looked like. On the next page are three possibilities. The graph of the gradient function does not give any indication of the actual Y values of the original graph, only its general shape. We cannot know where it crossed the axes for example.

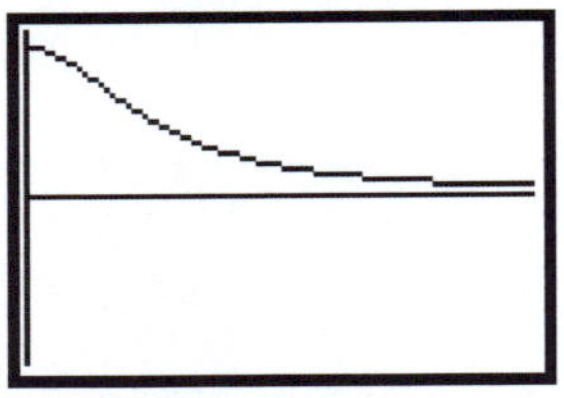
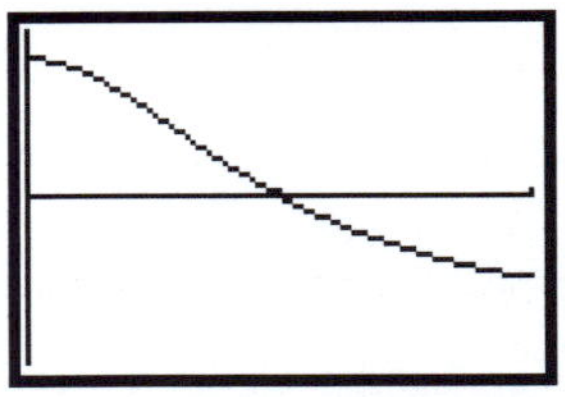
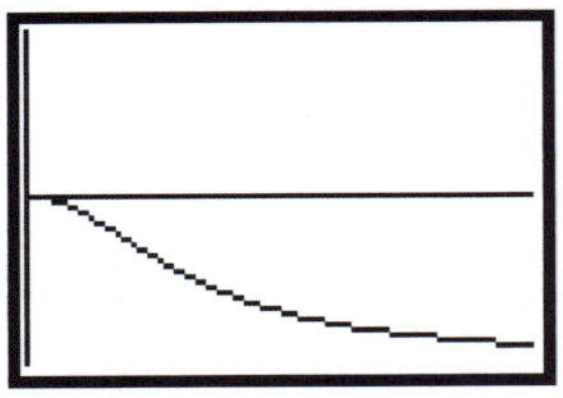

If you are interested you can see what function was actually used in this case by turning to the solutions at the back of this book. However, the point here is that there is a whole range of similar functions, any of which could have given rise to the gradient function.

(2) Working backwards

For each of the graphs of gradient functions shown below sketch a possible graph of the function stored in Y1.

In each case the window settings shown on the right have been used.

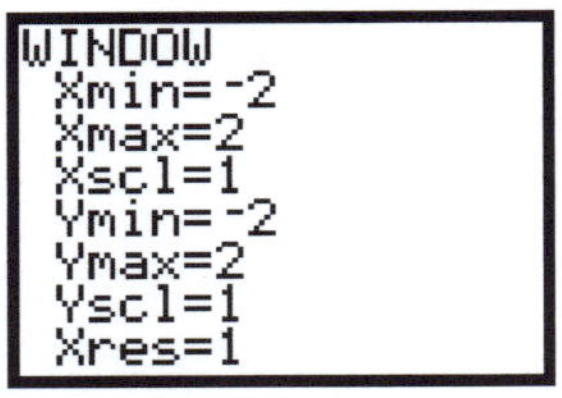

(a)

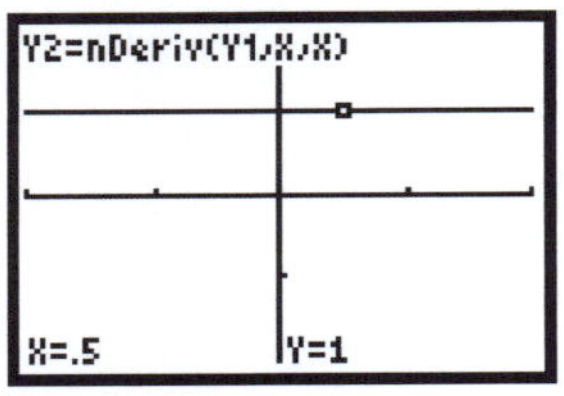

(b)

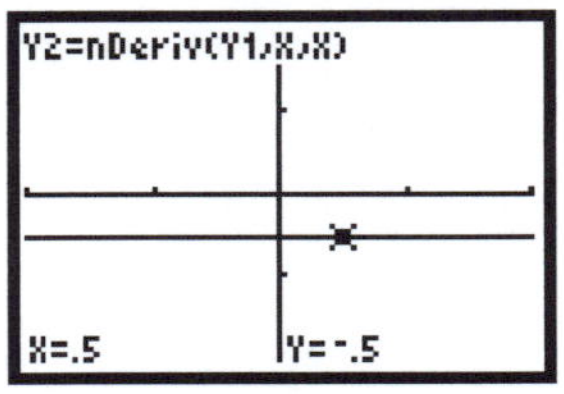

(c)

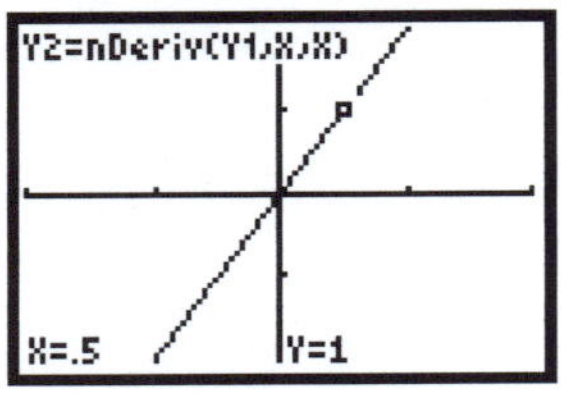

(d)

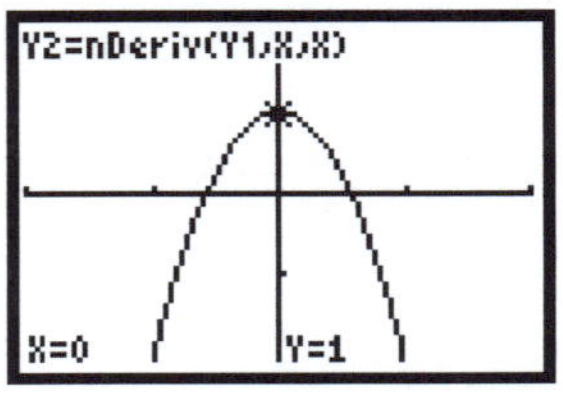

(3) Exploring MATH, option 9

The calculator's MATH menu contains a function that is closely related to what you have been doing in Activities 1 and 2. Option 9, **fnInt**, is in some senses the opposite of option 8, **nDeriv**.

Press [Y=] and enter the functions shown here. Notice that **fnInt** requires four arguments, which will be explained in a later unit.

For the moment you can interpret **fnInt(Y1,X,0,X)** as meaning an instruction to the calculator to produce a function whose gradient function is stored in Y1. So here it must find a function whose gradient is always $^{-}0.5$.

Choose the **ZDecimal** window settings to display the two graphs and compare the result with Activity 2(b).

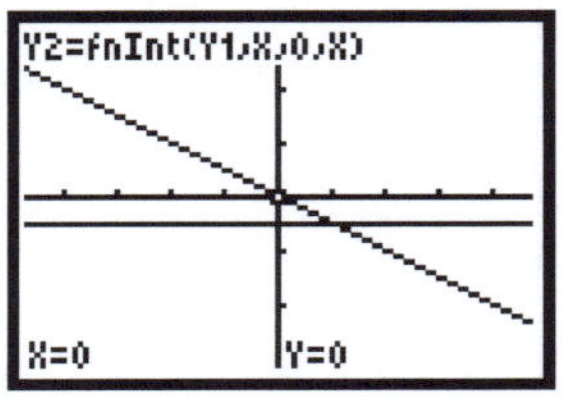

(4) Speeding up fnInt

You are sure to have noticed how slowly the calculator draws the line representing **fnInt(Y1,X,0,X)**. This is because it needs to do a *very* time-consuming calculation for every value of X from $^{-}4.7$ to 4.7 in steps of 0.1, that is for every pixel on the X-axis.

You can speed things up by changing the value of **Xres** in the Window settings. If, for example, you make **Xres=5** the calculator will evaluate the function only at every 5th pixel on the X-axis. It connects the points whose coordinates it has calculated with straight-line segments.

This makes no difference to the straight-line graphs you are graphing at the moment, but for curves a rather jagged graph will result. Notice also that changing **Xres** from the default value of 1 makes a difference to the way Trace operates.

The maximum possible value of **Xres** is 8.

(5) More antiderivatives

Leaving **Y2=fnInt(Y1,X,0,X)**, change Y1 to each of the following simple functions in turn and see an antiderivative function produced by the calculator. In each case, when the graphs are drawn, remind yourself that Y1 represents the gradient function of the function in Y2.

(a) **Y1=1** (b) **Y1=2.5** (c) **Y1=X** (d) **Y1=1–X²**

(6) There and back again

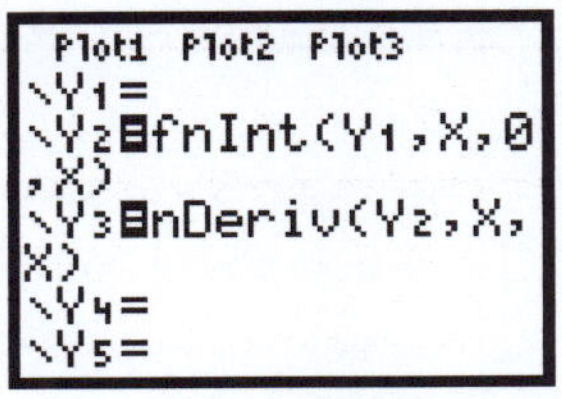

Set up your Y= screen to look like this and enter any function you like as Y1. How many curves would you expect to see on the graphing screen?

Since Y3 is the gradient of Y2, and Y2 is a function whose gradient is Y1, it is not surprising that Y1 and Y3 end up as the same function.

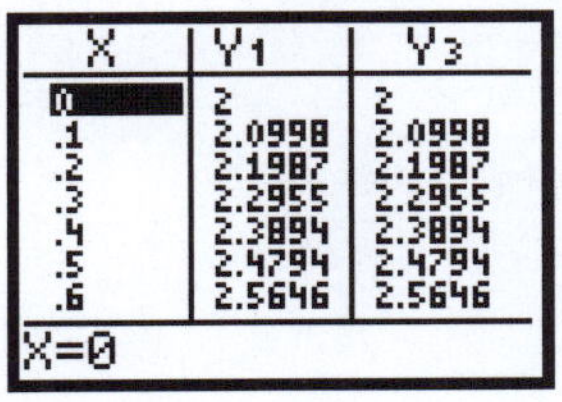

Another way to check this is to use the calculator's table facility, perhaps with Y2 deselected, as shown here. Try this with whatever function you have stored in Y1.

(7) Multiple antiderivatives

Look back to the end of Activity 1 and recall that when you work back from a gradient function, it is not possible to be totally sure what the original function was – there are actually lots of possibilities. This is why in Activity 5 we used the word *an* rather than *the* when asking you to "see an antiderivative function produced by the calculator."

Think about this a little more carefully by considering the gradients of $y = x^2-2$, $y = x^2-1$, $y = x^2$, $y = x^2+1$, and $y = x^2+2$.

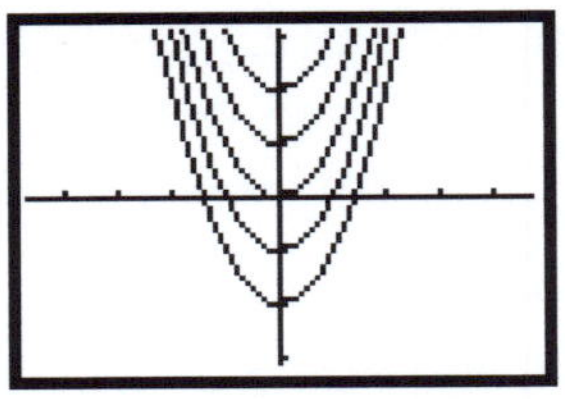

Their graphs are shown here.

Can you see that the gradients of all these curves are identical across all values of X?

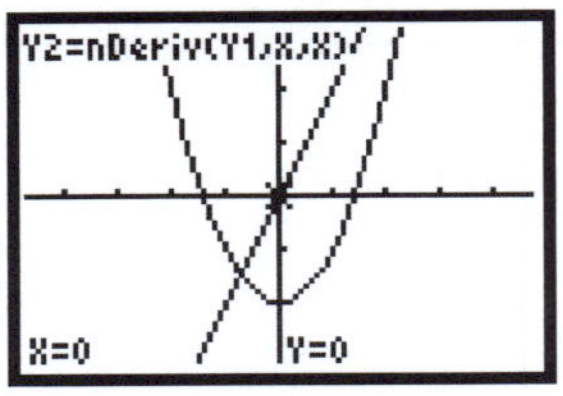

Try plotting any one of these curves and its derivative and you always get the same straight line.

Also, from the results you found in Unit 8 you will know that:

if $y = x^2-2$ then $\dfrac{dy}{dx} = 2x$,

if $y = x^2-1$ then $\dfrac{dy}{dx} = 2x$,

if $y = x^2$ then $\dfrac{dy}{dx} = 2x$ and so on.

Or in general:

if $y = x^2 + c$ then $\dfrac{dy}{dx} = 2x$ where c is any constant.

It follows that the antiderivative of $2x$ could be $x^2 -2$, $x^2 -1$, x^2 or in general $x^2 + c$.

An alternative term for *antiderivative* is *indefinite integral.* This name stresses that one cannot be definite about the function: the part of the function that involves x can be determined but any constant value, c, can also be a part of it.

(8) A notation for antiderivatives

Rather than writing "the antiderivative of $2x$", the notation $\int 2x\,dx$ is used.

Here $\int$ is called the integral sign and dx is used as a reminder that it is linked to differentiation with respect to x.

So in full $\int 2x\,dx = x^2 + c$

and this is read as "the indefinite integral of $2x$ with respect to x equals x^2 plus a constant of integration, c."

(9) A general rule for antiderivatives

At the end of Unit 8 there was a general rule for derivatives:

If $y = x^n$ then $\frac{dy}{dx} = nx^{n-1}$.

So if $y = x^{n+1}$ then $\frac{dy}{dx} = (n+1)x^n$.

It follows then that $\int (n+1)x^n\,dx = x^{n+1}$ + some constant

and dividing by n+1 gives $\int x^n\,dx = \frac{x^{n+1}}{n+1}$ + some constant.

And this is a general rule for antiderivatives, or for indefinite integrals.

Some people like to remember this using this slogan.

(10) Testing the rule for n=2

The general rule predicts that $\int x^2\,dx = \frac{x^3}{3} + c$.

It is possible to check this result using some of the techniques developed in earlier units.

Carry out the following steps:

Enter **Y1=X²** and **Y2=fnInt(Y1,X,0,X)**.

Plot the graphs with **Xmin=⁻2** and **Xmax=2** window settings.

Run program **SCATTER** (see Unit 7) to produce a scatterplot of some of the points lying on Y2.

Carry out a cubic regression calculation on the data in L1 and L2.

Do the calculated results confirm that $\int x^2\,dx = \frac{x^3}{3} + c$?

Brain stretcher **Exceptional *n***

Does the general rule for antiderivatives, $\int x^n\,dx = \frac{x^{n+1}}{n+1} + c$, hold for all integer values of n? What if $n = 1$, $n = 0$, $n = {}^-1$...?

(11) Integration practice

Use the general rule to calculate the following indefinite integrals. In each case check the result using the calculator. One way to do this is to enter in Y3 the function predicted by the general rule and then to compare the graphs or table values of Y2 and Y3. Alternatively you could use the method of Activity 10 with an appropriate regression calculation as indicated.

(a) $\int(4x^2 + 3x + 1)\,dx$ (If using regression, use **CubicReg**.)

(b) $\int(4x^3 - 2x)\,dx$ (If using regression, use **QuartReg**.)

(c) $\int\sqrt{x}\,dx$ (or $\int x^{\frac{1}{2}}\,dx$) (**Xres=8** is recommended. If using regression, use **PowerReg** and window settings **Xmin=0.1** and **Xmax=3**.)

Brain stretcher **There and not always back again**

In Activity 6 you saw that if you integrate and then differentiate you end up with the function you started with. However, the same is not always true if you differentiate and then integrate.
Try using various functions in Y1 and see if you can find any where the graphs of Y1 and Y3 are not the same. Can you explain why?

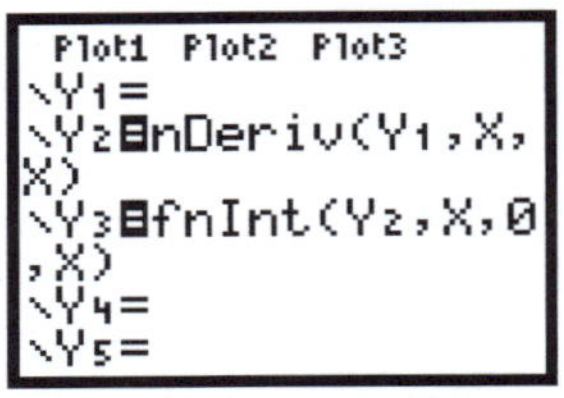

So finally ...

In this unit you have met the terms antiderivative *and* indefinite integral *(which mean very much the same thing) and also, $\int$, the integral sign.*

◊ *What sort of function would you expect to be produced by finding the indefinite integrals of the following:*

a constant,

a linear function,

a quadratic function,

a cubic function?

◊ *Can you explain in your own words why indefinite integrals require a constant of integration, c?*

Unit 12 Definite integrals

In Unit 10 you saw that antiderivatives (or integrals) are in some senses the inverses of derivatives and you used the function **fnInt(Y1, X, 0, X)** on the calculator's Y= screen to draw graphs of these integral functions.

Also in Unit 10 you met the use of the integral sign, ∫, to stand for the indefinite integral. This sign also appears in one of the calculator's menus and you will now be asked to investigate how it works. This will lead to an understanding of definite (as opposed to indefinite) integrals.

(1) Investigating ∫f(x)dx

Press [2nd] [CALC] to see the CALCULATE menu. You used option 6 in earlier units to calculate the gradient of functions plotted on the graphing screen. Now you are going to investigate option 7, **∫f(x)dx**. In order to do this, first set up the Format, Window and Y= screens as shown below.

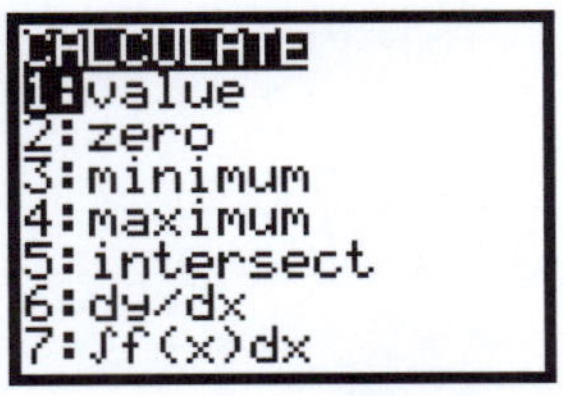

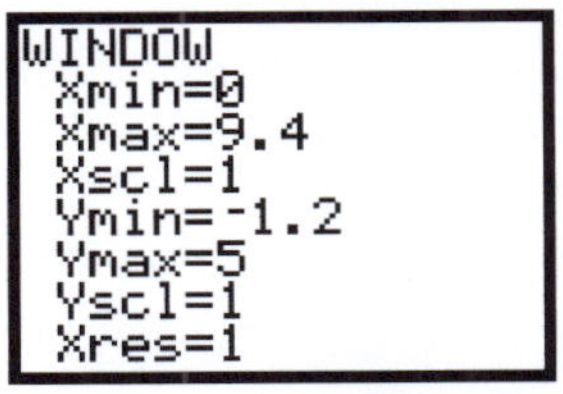

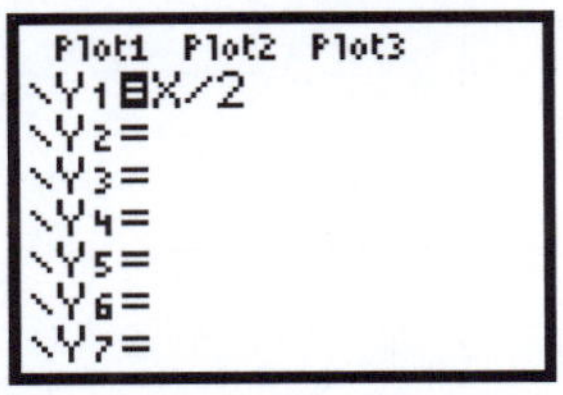

Press [GRAPH] to see the straight-line graph of the function $y = x/2$.

Now press [2nd] [CALC] **7** and try out the **∫f(x)dx** command. When asked for **Lower Limit**, use the left or right cursor keys and press [ENTER] to confirm the position (or alternatively, enter a numerical X value.) Enter an **Upper Limit** in a similar way.

Try to work out for yourself what the calculator is doing and how the calculated value of **∫f(x)dx** is produced. It may not be immediately obvious, but try it several times before going on to Activity 2.

(2) Systematic investigation

(a) As you saw in Activity 1, the command **∫f(x)dx** not only produces a calculated value: it also shades part of the graphing screen – but what is the connection between the two?

Use **ClrDraw** and then **∫f(x)dx** again, using X=0 as the lower limit and X=1 as the upper limit. Write down the value of **∫f(x)dx**. The points produced by the **GridOn** command enable you to imagine the screen covered with a square grid, with the area of each square equal to 1 square unit.

(b) Repeat part (a) several times, on each occasion using X=0 for the lower limit and the values X=2, X=3, X=4... for the upper limit.

Record your results in a table like this.

Lower limit	0	0	0	0	0	0	0	0	0
Upper limit	1	2	3	4	5	6			
∫f(x)dx									

(c) Can you spot the pattern between the values of the upper limit and of **∫f(x)dx**? What would you expect the value of **∫f(x)dx** to be, if the upper limit were 10, or 2.5, or *N*?

(d) From your work in Unit 10 you should be able write down the antiderivative (that is the indefinite integral) of the function $y = x/2$. How does this relate to your final answer in part (c)?

(e) Write a sentence or two summarizing what you know about the calculator's **∫f(x)dx** command.

(3) Changing the lower limit

Now consider what happens if you use **∫f(x)dx** again applied to the function $y = x/2$, but this time using something other than X=0 as the lower limit. You may be able to predict the outcome of some of the following – so carry out a "guess before you press" approach for each of the following.

(a) Find the value of **∫f(x)dx** for lower limit = 5, upper limit = 6.

(b) Find the value of **∫f(x)dx** for lower limit = 2, upper limit = 6.

(c) Find the value of **∫f(x)dx** for lower limit = 4, upper limit = 4.5.

(4) The area of a trapezium

There is a simple formula for calculating the area of a trapezium, such as those you saw in Activity 3. Expressed in words it goes as follows:

the area of a trapezium is
the mean length of the parallel sides
times the perpendicular distance between them.

Look carefully at the diagrams below – the first depicts a particular trapezium, corresponding to the one in Activity 3(b), and the second depicts a general trapezium drawn by the **∫f(x)dx** command, using lower limit, X = A and upper limit, X=B.

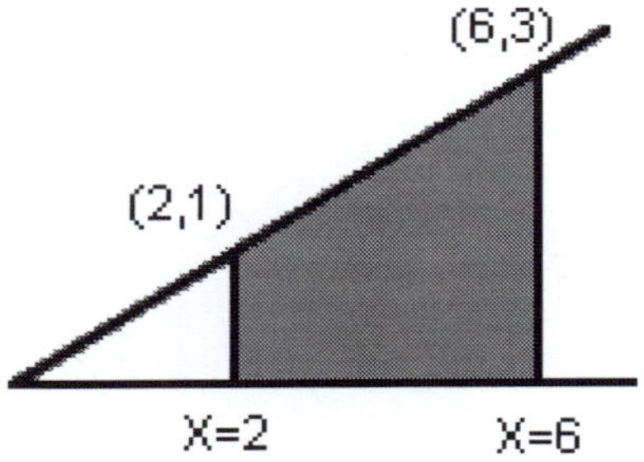

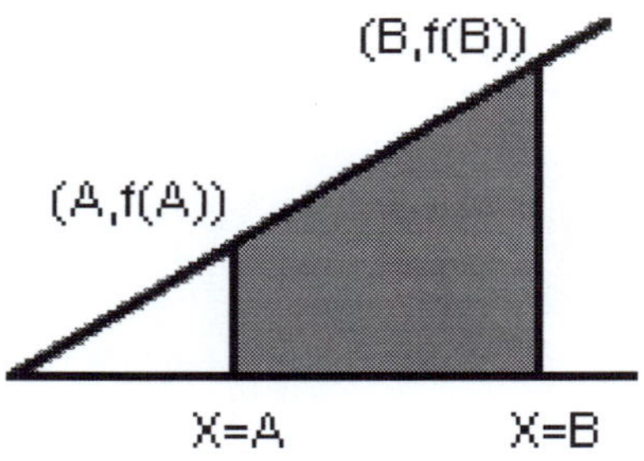

Area of trapezium = (mean of parallel sides) x (distance between)

$$= \frac{1+3}{2} \times (6-2) = 8 \qquad\qquad = \frac{f(A)+f(B)}{2} \times (B-A)$$

Check your answers to Activity 3(a) and (c), using the formula for the area of a trapezium. You will be using this formula again in Unit 12.

(5) Other functions, other areas

So far you have seen that applying the calculator command **∫f(x)dx** to the particular function $y = x/2$ produces the area beneath the graph between a lower limit such as X=*A* and an upper limit such as X=*B*. But is the same true for other functions?

(a) Enter **Y1= 2X/3**. Draw vertical lines at X=3 and X=6. (Option 4 in the DRAW menu). Calculate the area between the lines and beneath the graph, first by counting squares, then by using the formula for the area of a trapezium and finally using **∫f(x)dx**.

(b) Enter **Y1= X/4+1**. Calculate the area beneath this graph between X=1.5 and X=7.2, by using the formula for the area of a trapezium. Now use **∫f(x)dx** to check your calculation.

(c) In parts (a) and (b) the functions had straight-line graphs so it was relatively easy to calculate the area beneath the graph using the formula for the area of a trapezium. But what if the function has a curved graph? Enter **Y1= X²/15**. Consider the area beneath this graph between X=4 and X=8. Clearly this area is not a trapezium but you could still make an estimate for the area by counting squares. Do this and then check your estimate using **∫f(x)dx**.

(6) Integrals and areas – some theory

The results of Activity 5 suggest that the calculator command **∫f(x)dx** can be used to calculate areas under graphs generally. Here the calculator is drawing upon a well-established property of integrals, which can be expressed formally as follows.

> For any function y=f(x), the area bounded by the curve, the x-axis and the lines x=A and x=B can be determined by calculating the difference between the values of the integral function at x=A and x=B. This is known as the **definite** integral from A to B of f(x). A and B are known as the limits of integration and the notation used is $\int_A^B f(x)dx$.

So, for example to calculate the area of Activity 5(c) without using the calculator you would proceed as follows:

The area beneath the graph of $y = x^2/15$ between x=4 and x=8 is $\int_4^8 \frac{x^2}{15}dx$.

Using the general rule for integration established in Unit 10: $\int \frac{x^2}{15}dx = \frac{x^3}{45} + c$

When X=4 the value of the integral is 1.422 + c (to 3 d.p).

When X=8 the value of the integral is 11.378 + c (to 3 d.p).

So the difference between these two values, 9.956, gives the required area to 3 d.p. Notice that the constant of integration cancels out.

This process is normally written more concisely like this:

Required area $\int_4^8 \frac{x^2}{15}dx = \left[\frac{x^3}{45}\right]_4^8 = \frac{8^3}{45} - \frac{4^3}{45} = 11.378 - 1.422 = 9.956$ to 3 d.p.

Recall that this is the value obtained by using the **∫f(x)dx** command in Activity 5(c).

(7) Practising definite integration

Here are some definite integrals to evaluate. In each case work them out first without using the **∫f(x)dx** calculator command, setting out your working using the concise notation used at the end of Activity 6. Then check your answer by entering the relevant function on the Y= screen of the calculator and using **∫f(x)dx**.

(a) $\displaystyle\int_1^2 \frac{x^2}{15}dx$ (b) $\displaystyle\int_2^4 (\frac{x^2}{15} + \frac{x}{4})dx$

(c) $\displaystyle\int_0^8 (2x - \frac{x^2}{4})dx$ (d) $\displaystyle\int_1^5 x^{\frac{1}{2}}dx$

(8) Keep it even simpler

There is an even quicker, simpler way of evaluating definite integrals on the calculator. This method uses the **fnInt** command, which you first met in Unit 10. Look back now at Activity 3 in Unit 10 where you entered **Y2=fnInt(Y1,X,0,X)** to produce the graph of the integral of the function stored in Y1.

However, a similar command can be used directly on the home screen to calculate definite integrals of any function, using the last two arguments as the limits of integration. For example, the screen shown on the right is evaluating the definite integral

$$\int_4^5 (x-2)dx.$$

```
fnInt(X-2,X,4,5)
                2.5
```

Evaluate the four definite integrals in Activity 7 using this method.

Before moving on, select **GridOff** on the FORMAT menu.

So finally ...

Calculation of areas under curves is one of the most important applications of integration and this unit and the next deal with this important idea.

◊ *In Activities 1 to 4 you saw that the* **∫f(x)dx** *calculator command produces the area under a straight-line graph and checked this by counting squares or calculating areas of trapeziums. (You'll be using this technique again in Unit 13).*

◊ *You saw in Activities 5 and 6 that* **∫f(x)dx** *can be used to calculate areas of functions generally and that the command evaluates what is known as a* definite integral.

◊ *In the previous unit you saw how to derive the* indefinite *integrals of many functions and in Activities 6 and 7 you used those techniques of integration to evaluate some definite integrals, checking the results using* **∫f(x)dx** *and also the* **fnInt** *command.*

All this should have laid the foundation for more work on integration and area in the next unit, together with consideration of how the calculator carries out integration.

Unit 13 Integration and area

In Units 11 and 12 you used the calculator's commands **fnInt and ∫f(x)dx** to help understand integration. How they were used is summarised below.

Calculator command:	*Used on:*	*To give:*
fnInt	Y= screen	graphs of antiderivatives (or indefinite integrals)
fnInt	home screen	definite integrals
∫f(x)dx	graphing screen	definite integrals, in the form of areas beneath a graph

It is important to realise that members of the TI-83 family of calculators are not able to carry out algebraic manipulation and integrate a function directly to provide an *algebraic* result.

The calculator cannot tell you, for example, that $\int x^2\,dx = \frac{x^3}{3} + c$.

Rather, the calculator can carry out so-called **numerical integration**.

Long before calculators existed, various methods of carrying out numerical integration had been important. The reason for this is that it is not always possible to calculate indefinite integrals:

many functions (like $y=x^2$, $y=\sin(x)$, etc.), can be integrated using simple algebra;

many others can be integrated using advanced algebraic techniques (such as integration by parts, integration by substitution, etc.);

there remain other functions (such as $y=e^{-x^2}$) that cannot be integrated algebraically at all.

With such functions, some form of numerical integration is required.

One way of doing numerical integration is to calculate the area under a curve as accurately as possible and then to equate this to the required integral. This unit deals with one such traditional method for calculating areas. In addition there are activities that deal with more complicated areas under curves.

(1) The trapezium rule for numerical integration

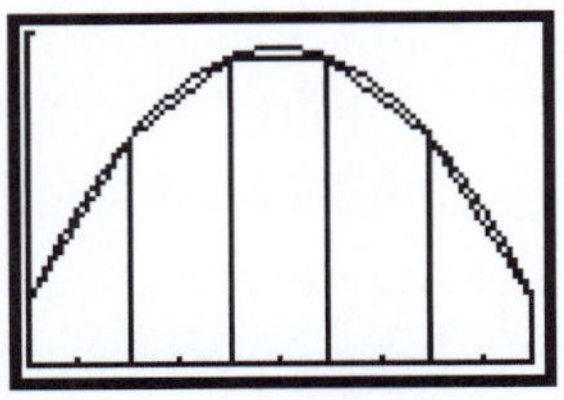

In Unit 12 you used the method of counting squares to estimate the area under a curve. A rather better approximate method of calculating the area under a curve involves dividing it into several trapeziums, calculating the area of each one and finding the sum. For example on the right, the area under a curve has been divided into five trapeziums of equal width. It would be possible to calculate the area of each trapezium using the formula from Unit 12, that is:

the area of the trapezium beneath function $y=f(x)$ between $x=A$ and $x=B$ is $\frac{f(A)+f(B)}{2} \times (B-A)$.

Since all the trapeziums have equal width, *B–A* is the same for each trapezium and this simplifies the calculation somewhat. Of course, the more trapeziums, the smaller their width and the more accurate the approximation to the true area under the curve. Such an approach is known as using the Trapezium Rule for numerical integration and it has been used for centuries to calculate numerical integrals.

(2) A program for the Trapezium Rule

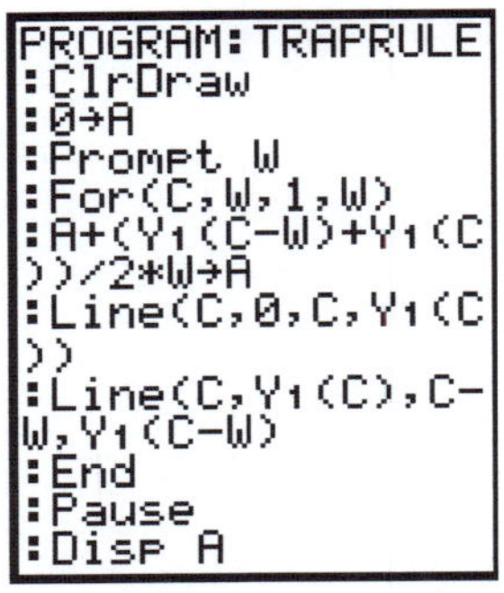

Here is a program that will calculate an approximation for the area under a curve between X=0 and X=1 where the function is stored in Y1. The program allows you to decide the width of the trapeziums (stored in W). It draws the trapeziums themselves, pauses until you press [ENTER] and then displays the approximation to the area (stored in A).

The series of screen displays below shows the program being used to calculate a series of approximations for the area under the curve **Y1=√(X)**.

First the Y= screen has been set up and an appropriate viewing window formed.

The program is executed and W, the width of the trapeziums, set to 0.5. This first approximation to the area under the curve uses just two trapeziums and clearly leaves a large section out.

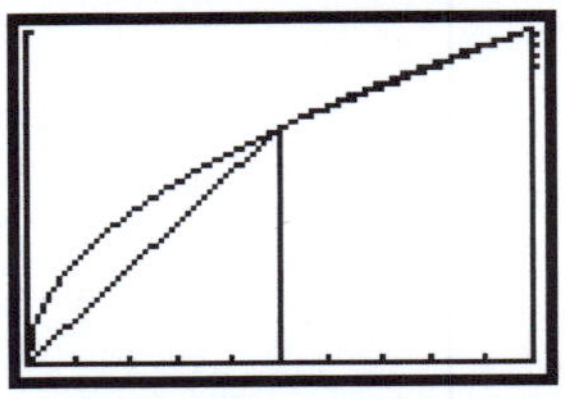

A better approximation can be achieved using W=0.2, giving 5 trapeziums.

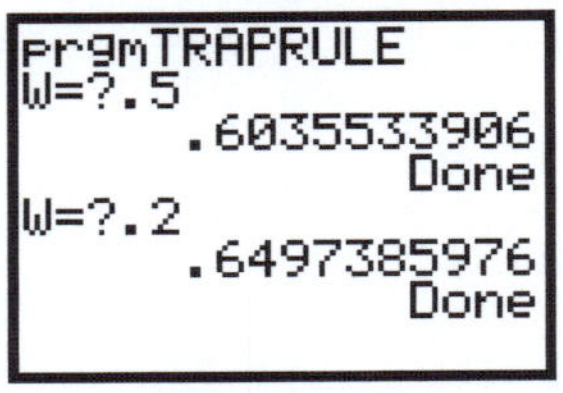

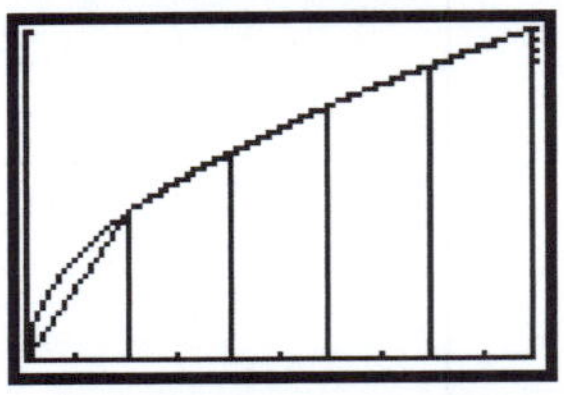

Reducing the value chosen for W increases the number of trapeziums and improves the approximations still further.

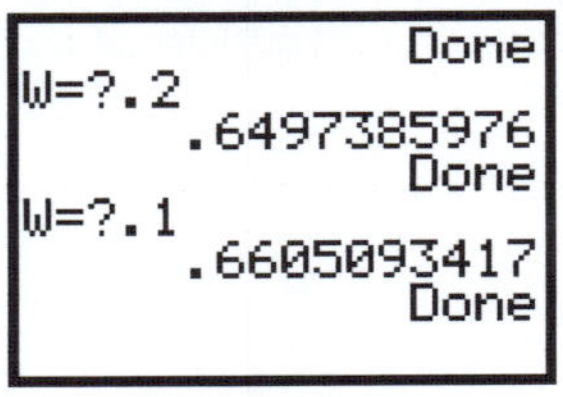

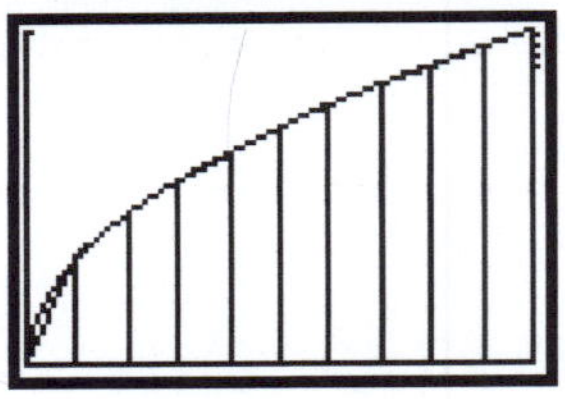

(a) Enter the program on you calculator. If you need help, the key presses are given in the solutions.

(b) Repeat the process shown above, to get increasingly accurate approximations of the numerical integral of $y=x^2$ between $x=0$ and $x=1$, that is of $\int_0^1 x^2 dx$.

(c) Use the program again to get better and better approximations of $\int_0^1 4x(1-x)dx$.

(3) A better approximation for definite integrals

Using the calculator's command **∫f(x)dx**, evaluate $\int_0^1 4x(1-x)dx$.

Compare this solution with your answer to Activity 2(d). It should be clear that the calculator does not use the Trapezium Rule to produce an approximation for definite integrals.

In fact the calculator uses a much faster routine known as the Gauss-Kronrod method. The mathematics lying behind this method is beyond the scope of this book. However, in the rest of this unit you are asked to use **∫f(x)dx** and **fnInt** to further your understanding of areas under graphs.

(4) A curious area

(a) Look at the screen display shown here.

It seems to show that $\int_1^3 (x-1)(x-2)(x-3)dx = 0$.

Plot the graph of $y = (x–1)(x–2)(x–3)$ on your calculator, choosing appropriate window settings. Does it seem likely that the area lying between this curve, the x-axis, x=1 and x=3 is 0?

Use **∫f(x)dx** to confirm the above result.

Can you explain what is going on?

(b) Now calculate the integral of $y = (x–1)(x–2)(x–3)$ between the limits x=1 and x=2, followed by the integral between the limits x=2 and x=3. Does this explain the result in part (a)?

(c) Evaluate $\int_1^3 (x-1)(x-2)(x-3)dx$ by:

expanding the brackets,

carrying out an indefinite integration using the rule developed in Unit 11,

applying the limits of integration x=1 and x=3.

(d) Clearly the area between the graph of $y = (x–1)(x–2)(x–3)$ and the x-axis between x=1 and x=3 is not zero. There is an area of 0.25 sq. units above the x-axis and an equal area of 0.25 sq.units below the x-axis. So the total area is 0.5 sq. units.

However $\int_1^3 (x-1)(x-2)(x-3)dx$ does indeed equal 0.

When calculating the area between a curve and the x-axis, care has to be taken if some of the area lies above and some below the axis. In such cases it is often best to treat each region separately.

(5) Which side of the x-axis?

In each part of this activity, the area is not the same as the corresponding definite integral. Use your calculator to help you find the areas between:

(a) $y=x^3$, the x-axis, $x=^-2$ and x=2.

(b) $y=x^3$, the x-axis, $x=^-1.2$ and $x=^-0.8$.

(c) $y=(x-1)^3$, the x-axis, x=0 and x=2.5.

(d) y=sin(x) the x-axis, x=0 and x=2π radians.

(6) A very complex area

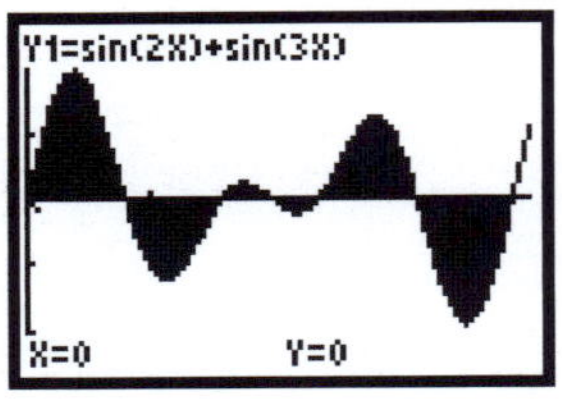

(a) Consider the total area shown shaded here.

The equation of the line is **Y1= sin(2X)+sin (3X)** and the area shown shaded was produced using **∫f(x)dx** with lower limit = 0 and upper limit = 2π. Try this for yourself and find the tiny value produced for **∫f(x)dx**. It is very close to zero, an unlikely total for the sum of the six shaded parts.

(b) How would you go about calculating the total shaded area?

Give some thought to this question before reading one possible strategy below.

The fact that **∫f(x)dx** is very close to zero, together with the shape of the graph, suggests that it might be symmetrical about the middle X value, that is X=π. If that were true you could find the sum of the three areas to the left of X=π and double the result.

However, you would need to find where the graph crosses the x-axis in order to find the areas of each of the three shaded sections to the left of X=π. This could be done using option 2 in the CALC menu, **zero**. You would need to find the two pertinent zero values, perhaps storing them in **A** and **B**.

Then you could use **∫f(x)dx** three times to calculate the three sub-areas between the curve, the x-axis, X=0 and X=π partitioned by the values of **A** and **B**. Each sub-area could be stored as it is calculated. You would then need to add the three stored areas together and finally, to get a value for the total of all six shaded areas, you would double your last answer.

If you were to carry out this process you would find that the area is 5.060 square units (to 3 d.p.). Phew!

(7) A sneaky shortcut for complex areas

Rather than the long haul of Activity 6 you may be pleased to know that there is an alternative method for such calculating such nasty complex areas. Since the problem is caused by having shaded areas both above and below the x-axis, why not use the **abs** function to produce a similar graph which lies entirely above the x-axis? Study the screen displays here to see this technique.

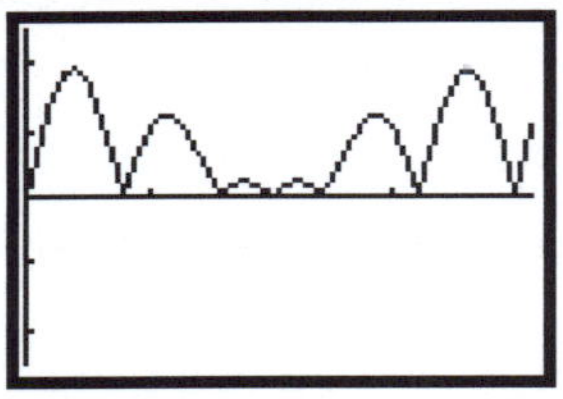
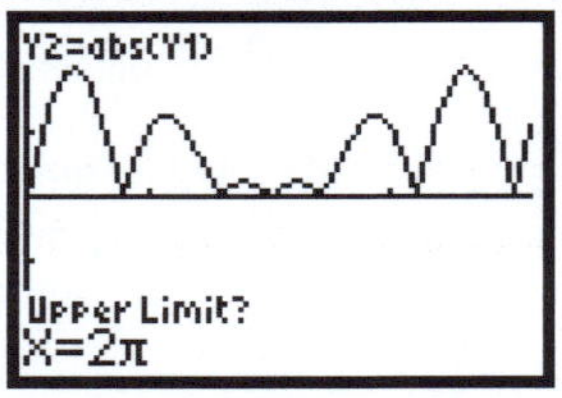

Unfortunately the calculator takes an *extremely* long time to produce the final graph and calculation – but it does mean you can take things easy while it's working its socks off!

(a) Try out this technique for yourself to produce the displays shown above.

(b) Choose one or more of the questions in Activity 5 and use this sneaky shortcut to check your previous results.

(8) The area between two curves

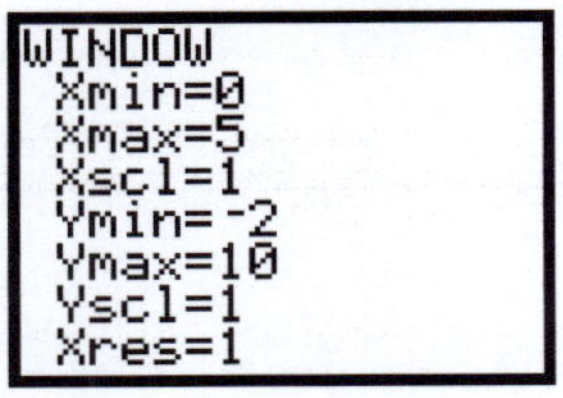

(a) On your calculator enter the functions **Y1=X²/2+2** and **Y2=2X+2** and draw their graphs using the window settings shown here. Decide how would you go about calculating the area between the two curves and X=1 and X=3?

(b) Calculate the area under the higher of the two curves and X=1 and X=3 using the **fnInt** command on the home screen. Now do the same for the lower of the two curves. Subtract the two areas to find the area *between* the two curves.

(c) A quicker method of finding the area between the two curves is to enter **fnInt(Y2–Y1,X,1,3)** on the home screen. Try entering it now. You should get the same answer as for part (b) because in general:

$$\int_A^B f(x)dx - \int_A^B g(x)dx = \int_A^B (f(x) - g(x))dx$$

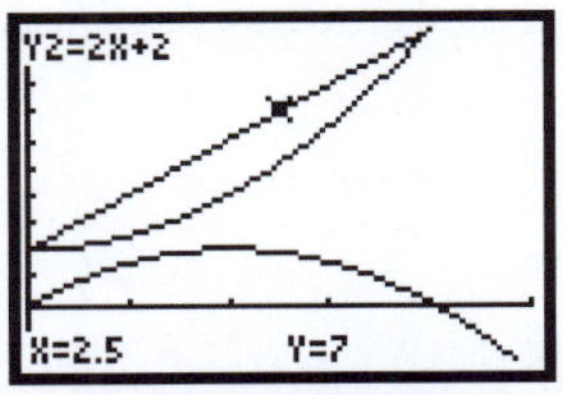

(d) To help you visualise why the definite integral of Y2–Y1 gives the required area, on the Y= screen enter **Y3=Y2–Y1** and plot the graphs of the 3 functions. Look carefully at how the three curves relate to each other, perhaps using [TRACE] [▼] [▼] to compare corresponding points.

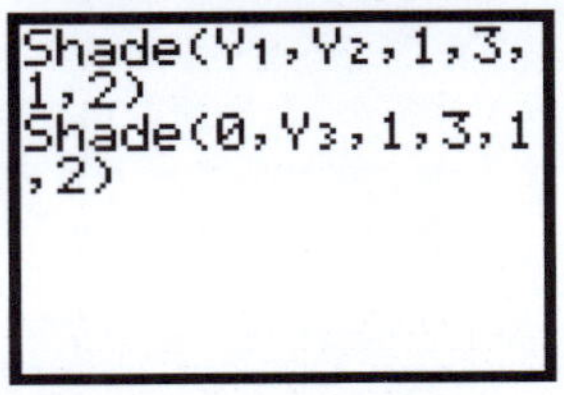

Finally, to see how the two areas relate directly to each other, use the **Shade** command in the DRAW menu as shown here. (See page 8-10 of the TI-83 Guidebook if you want to know more about the use of the **Shade** command.)

(e) The algebraic method for finding this area is as follows:

$$\text{Area} = \int_1^3 (2x+2)dx - \int_1^3 (\frac{x^2}{2}+2)dx = \int_1^3 (2x+2-(\frac{x^2}{2}+2))dx$$

$$= \int_1^3 (-\frac{x^2}{2}+2x)dx$$

$$= \left[-\frac{x^3}{6}+x^2\right]_1^3$$

= 3.667 sq. units (to 3 d.p).

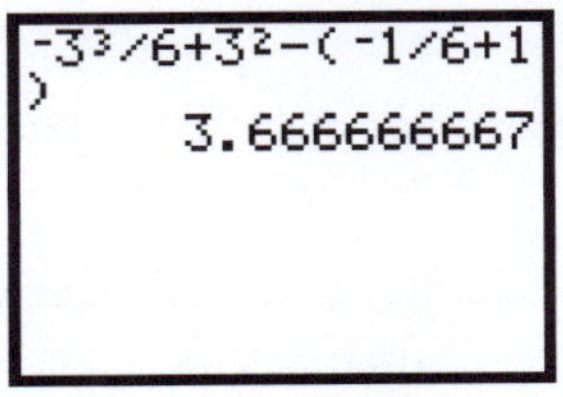

(9) Two crossing curves

(a) On your calculator enter the functions **Y1=X+2/3** and **Y2=X²–2** and draw their graphs using the **ZDecimal** window settings. Decide how would you go about calculating the area between the two curves.

(b) Here is a first quick graphical method. In this case the graph of Y1 is above that of Y2 for the area in question, so enter **Y3=Y1–Y2**. Recalling the methods of Activity 6, the required area is the area between Y3 and the x-axis, but in this case the actual crossing points are unknown. However, you can use **∫f(x)dx**, tracing along the graph of Y3 to enter approximate Lower and Upper limits.

(c) Now for a slightly longer, but more accurate, graphical method. First, find the two intersections of Y1 and Y2 using the **intersect** command of the CALC menu; store the resulting X values in S and T respectively. Then calculate **fnInt(Y1–Y2,X,S,T)**.

(d) Finally, here is a method based upon the traditional algebraic approach. The calculator is used only to carry out arithmetic – the graphing screen is not used at all.

The first step is to find the intersection of $y=x+2/3$ and $y=x^2-2$. Treat the two as simultaneous equations and the quadratic equation $x^2-x-8/3=0$ is formed. Solve this equation using the formula method and store the two solutions in S and T.

On the right is shown a short program for solving quadratic equations.

Next, integrate: $\int((x+\frac{2}{3})-(x^2-2))dx \quad =\int(^-x^2+x+\frac{8}{3})dx$

$$=\frac{-x^3}{3}+\frac{x^2}{2}+\frac{8x}{3}+c$$

Finally this expression needs to be evaluated between the limits of integration, S and T.

One way of doing this on the calculator is shown here.

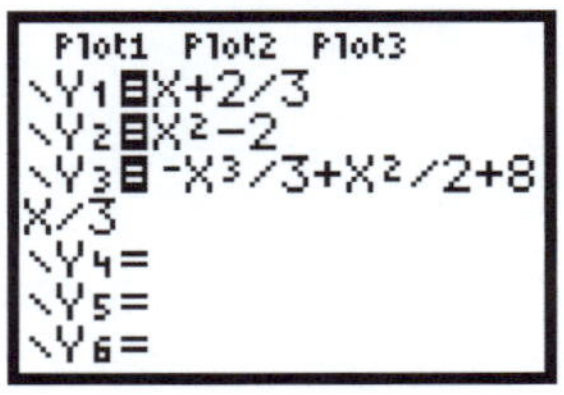

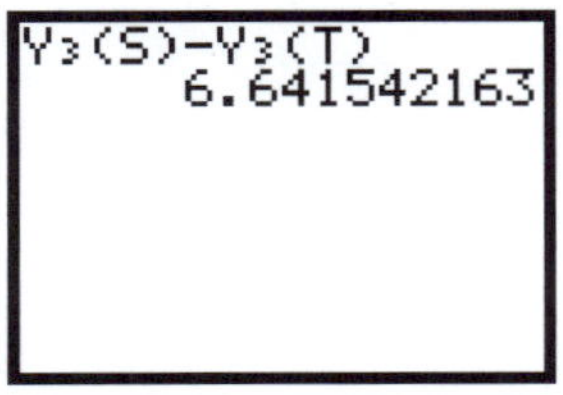

(10) Exercise area

As you have seen, there are various available methods to calculate area. Using whichever method you wish to practise, answer the following questions, correct to 3 d.p.

(a) What is the total area between $y=\sin(x)+\sin(2x)$ and the x-axis, between X=0 and X=2π?

(b) What is the total area enclosed by the curves $y=x^3$ and $y=\sqrt[3]{(x)}$?

(c) What is the area between the curve $y=1/(1-x)^2$ and the straight-line $y=4-X$?

(d) What is the area of the segment cut off the curve $y=5x-x^2$ by the line $y=6$?

So finally …

Is it correct to say that $\int_a^b f(x)dx$ *equals the area between the curve* y=*f*(x), *the x-axis and the lines* x=*a and* x=*b? … Always?*

Unit 14 Means, volumes and changing rates

One of the important applications of integration is to calculate the mean value of functions. By working through the particular scenario of Activities 1 to 3 you should get a good idea how the mean of a function can be calculated.

The later activities illustrate two other important applications of integration – to calculate volumes of revolution and to move from the rate of change of a variable back to the variable itself.

(1) That's a mean temperature

In a small community in the far north at the spring equinox, when days last 12 hours, T, the daytime temperature in °C, can be estimated from the formula $T = 2x - \frac{x^2}{6}$ where x represents the number of hours after sunrise. What is the mean daytime temperature?

(a) Enter the function on your calculator as **Y1=2X–X²/6**. Draw a graph with suitable window settings to get a representation of how the temperature varies during the day.

(b) One way to get an estimate for the mean temperature is to use the function to evaluate the temperature every hour and then to calculate the mean. There are various ways of doing this on the calculator but here is a single command, which will be easy to replicate and edit later.

It uses the **mean** and **seq** commands (from the LIST sub-menus) to produce a list of hourly temperatures and their average. Enter it on your calculator now by pressing:

```
mean(seq(Y1,X,0,
12,1))
```

[2nd] [LIST] [◄] **3** [2nd] [LIST] [►] **5** [VARS] [►] **1 1** [,] [X,T,Θ,n] [,] **0** [,] **12** [,] **1** [)] [)] [ENTER]

Does this seem a reasonable answer? Check on the graphing screen.

(c) A more accurate mean temperature can be found by evaluating the temperature every quarter of an hour. To do this simply press [2nd] [ENTRY] and change the final parameter to 1/4.

(d) Get even better answers by using time intervals of 5 minutes (1/12 of an hour) and 1 minute (1/60 of an hour). This last calculation takes a long time because a list of over 720 temperatures is being generated. If it were possible to use even shorter time intervals designated as δx, what mean temperature would you expect? Using the terminology used in previous units you could say: in the limit, as δx tends to zero, the mean temperature tends to ...what?

(e) On the Y= screen enter **Y2=4**. Check on the graphing screen to ensure that this seems a reasonable value for the mean temperature.

Brain stretcher **Mean or median?**

What difference does it make if you use the median, rather than the mean in Activity 1?
Repeat the exercise and plot Y2=median value rather than Y2=mean value.
Use **intersect**, option 5 in the CALC menu, to find where Y1 and Y2 cross.
Can you explain the values that appear?

(2) What's the meaning of mean – in terms of area?

(a) Look again at the graphing screen showing the functions **Y1=2X–X²/6** and **Y2=4**.

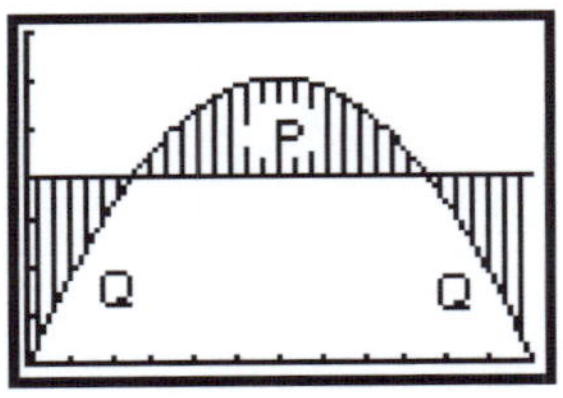

The effect of calculating and drawing the mean is to make shaded area P equal to the sum of the two shaded areas Q: in terms of temperature all the parts of temperatures above 4°C will balance the sum of all the parts below 4°C.

Another way of stating this is that the area under the temperature function **Y1=2X–X²/6** between x=0 and x=12 is equal to the area under the mean temperature **Y2=4**. Check this statement, using either of the calculator commands **fnInt** or **∫f(x)dx**.

(b) Since the two areas are equal, and since the area under the mean is easy to calculate because it is a rectangle, it follows that in this case:

Area under the curve, $\int_0^{12}(2x-\frac{x^2}{6})dx$ = mean × (12 – 0).

Therefore the mean of the function between x=0 and x=12 is $\frac{1}{12-0}\int_0^{12}(2x-\frac{x^2}{6})dx$.

Or, in general,

the mean of a function y=f(x) between the values x=A and x=B is $\frac{1}{B-A}\int_A^B f(x)dx$.

(3) A program for the mean of a function

Here is a very short program to calculate the mean of the function stored in Y1, between the values x=A and x=B.

(a) Create the program in your calculator (the necessary key presses are in the solutions).

(b) Test the program by calculating the mean temperature between x=0 and x=12, i.e. during the twelve hours of daylight.

(c) Use the program to calculate the mean temperature during the middle 4 hours of daylight.

(4) Meaningful practice

(a) A scientist observed that over a period of 4 hours the number of bacteria in a sample increased exponentially according to the rule $N = 5000 \times 2^t$ (where t represents the number of hours). Enter an appropriate function on the Y= screen and draw a graph to represent the population growth. Calculate the mean number of bacteria present during the 4 hour period and compare it with the actual number present when t=2.

(b) Find the mean value of sin(*x*) over the range *x*=0 to $x=\frac{\pi}{2}$ and also the range $x=\frac{\pi}{4}$ to $x=\frac{3\pi}{4}$.

(5) Solids of revolution

Another use of definite integration is to calculate volumes.

If the area under the curve *y*=f(*x*) between *x*=A and *x*=B is rotated through 2π radians about the x axis it sweeps out a so-called solid of revolution whose volume is given as:

$$\text{volume of revolution} = \pi\int_A^B y^2 dx.$$

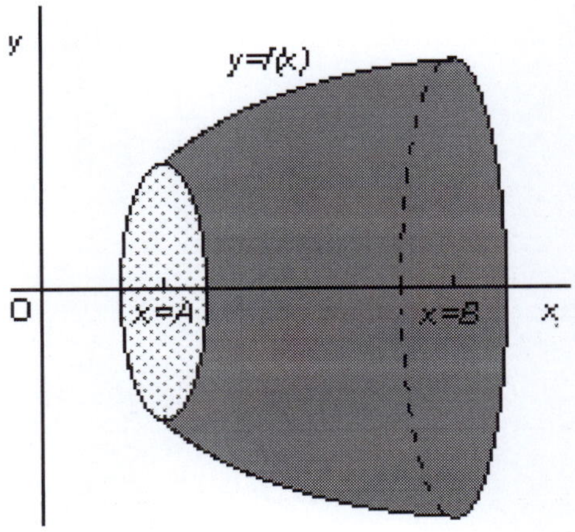

(a) Follow the steps below on your calculator to find the volume of the solid produced when the curve *y*=*x*(1–*x*/8) between *x*=1 and *x*=3 is rotated about the x-axis.

Enter the function **Y1=X(1–X/8)** and display the curve using the **ZDecimal** window settings.

Shade the area below the curve and between X=1 and X=3, using **∫f(x)dx** from the CALC menu and try to visualise the solid of revolution.

On the home screen enter **fnInt(Y1²,X,1,3)** and multiply by π to calculate the volume.

(b) Use **fnInt** to calculate the volume of the solid of revolution formed by rotating through 2π radians the area under the curve $y=\sqrt[3]{x}$ between *x*=10 and *x*=20.

(6) A sphere, a cone and a doughnut

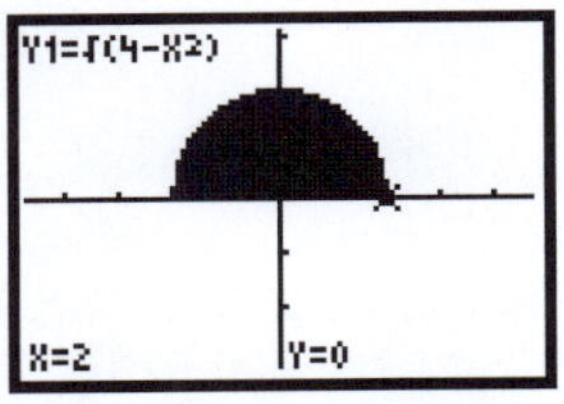

(a) A sphere can be formed by rotating a semi-circle about the x-axis. Enter the function **Y1=√(4–X²)** and use **fnInt** to calculate the volume of the corresponding sphere.

Check the result using the standard formula for the volume of a sphere: $V=\frac{4}{3}\pi r^3$.

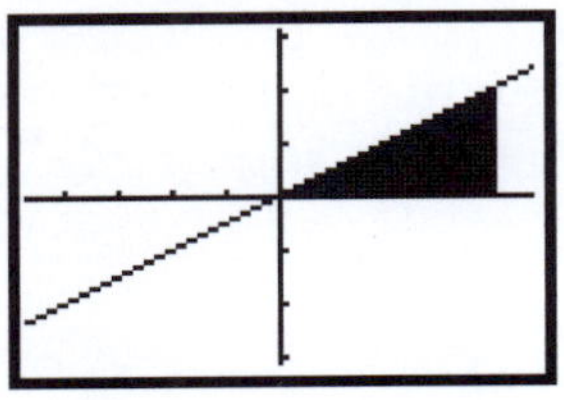

(b) A cone can be formed by rotating about the x-axis the area under a straight line through the origin.

The standard formula for the volume of a cone is $V=\frac{1}{3}\pi r^2 h$.

Check the standard formula by entering any appropriate function in Y1 and using **fnInt** to calculate the volume.

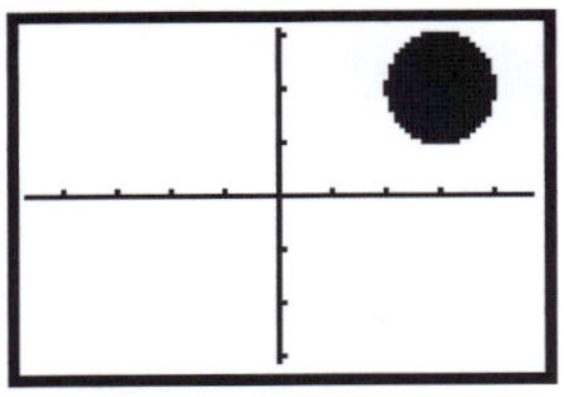

(c) A circle can be formed using two equations, one for the upper semicircle and one for the lower. Produce such a circle with radius 1 which has been shifted three units to the right and one unit up by entering:

Y1=√(1–(X–3)²)+2 and **Y2=⁻√(1–(X–3)²)+2**

Now imagine the shape that would be produced if the area inside the circle were to be rotated about the x-axis.

In mathematics this shape is known as a torus: elsewhere it is described as a doughnut!

Using **fnInt** it is possible to find the volume of the doughnut. Can you do it?

(7) Integrating rates of change

Integration is frequently useful to help you deal with rates of change. As you saw in Unit 10, integration is the reverse process of differentiation, so if a variable is known to be changing at a particular rate (i.e. its derivative with respect to time is known), you can calculate how much the variable changes between particular times by evaluating the definite integral of the rate of change.

Follow through this worked example and then, in the next activity, try some for yourself.

Suppose that a new company is increasing its production at a rate of $100e^{0.5x}$ units per month where *x* represents the time measured in months. How much will production increase over the first and second three-month periods of the coming year?

Notice that you know the *rate of change* of production, so if y represents the number of units produced and x the time in months, you know that $\frac{dy}{dx} = 100e^{0.5x}$. In order to calculate the actual change of production you need to evaluate $\int_0^3 100e^{0.5x}\,dx$ and $\int_3^6 100e^{0.5x}\,dx$.

It would be possible to enter **Y1= 100e^(0.5X)** and then use **fnInt(Y1,X,0,3)** and **fnInt(Y1,X,3,6)**, but the screen display here shows a more direct way of evaluating these definite integrals on the calculator.

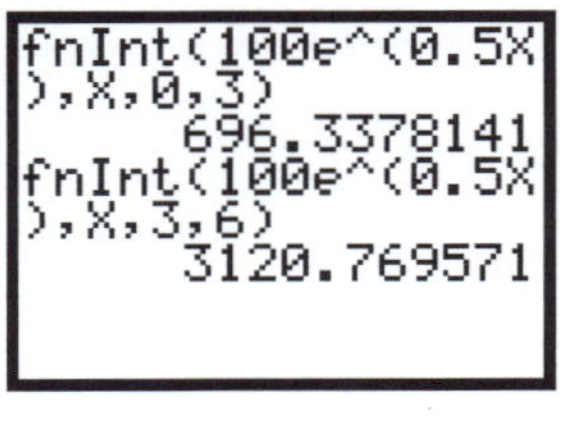

So the production will increase by 696 units in the next three months and by 3121 units in the three months after that.

(8) Fever

Doctors are considering using a drug to control a patient's fever. It is claimed that the drug will produce temperature changes at a rate of $0.4T-T^2/4$ degrees C per hour, with T measured in hours. Calculate the expected change in temperature (correct to 2 d.p) during the first, second and third hours of treatment.

(9) From velocity to displacement

(a) A particle moves in a straight line so that, X seconds after passing a fixed point A on the line, its velocity is $3X^2+2X+4$ m/s. (Recall that velocity is the rate of change of displacement over time.) The particle passes point B after 0.6 seconds and point C after 1.1 seconds. Use the **fnInt** command to calculate the distance apart (correct to 2 d.p) of the points A, B and C.

(b) Check your answers by drawing a displacement-time graph (that is with displacement on the vertical axis and time on the horizontal) showing the motion of the particle over the first 1.2 seconds of its motion.

So finally ...

◊ *In conventional textbooks on calculus you will find many other examples that involve evaluating definite integrals. In such circumstances, if the function to be integrated is known, your calculator will always be able to produce the required value, using its capacity to carry out a numerical integration.*

◊ *If in an examination you meet the following question, your calculator can answer it.*

Evaluate $\int_0^1 (\sin x + \cos x)dx$.

However, it is worth remembering that this calculator does not perform algebraic integration. If the question is worded like this, the calculator can only be used for the final part:

Calculate $\int (\sin x + \cos x)dx$ and use it to evaluate $\int_0^1 (\sin x + \cos x)dx$.

And absolutely finally ...

We hope that, having worked through this book, you have achieved three things:

gained a good grasp of the Calculus commands of your calculator,

gained a deeper understanding of the meaning and processes of differentiation and integration,

had some fun in the process!

Solutions

Unit 1 Gradient of a straight line

(3) (a) It is quite difficult to put the lines in order of steepness by eye. In Activity 5 you will use another method and be able to check your attempt to estimate steepness.

(4) Notice that the mathematical definition of gradient (the distance up divided by the horizontal distance along) differs from that often used to indicate the steepness of roads (the distance up divided by the distance along the road.)

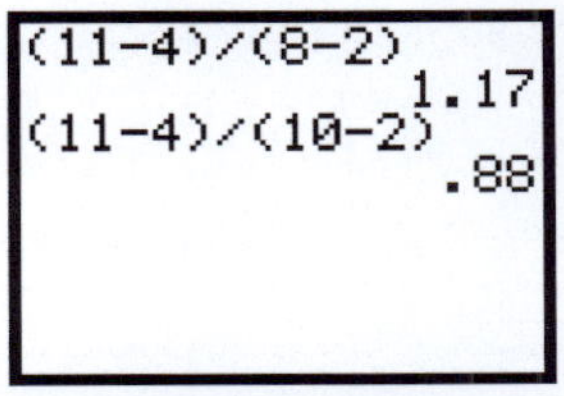

(b) To edit the previous command you need to press:

[2nd] [ENTRY] [◄] [◄] [◄] [◄] [DEL] [2nd] [INS] **10** [ENTER]

(5) If you used the lines drawn in Activity 2 you should have the calculations shown here, enabling you to place the lines in order of steepness.

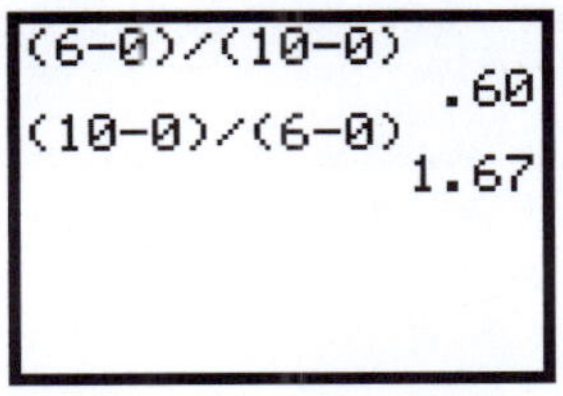

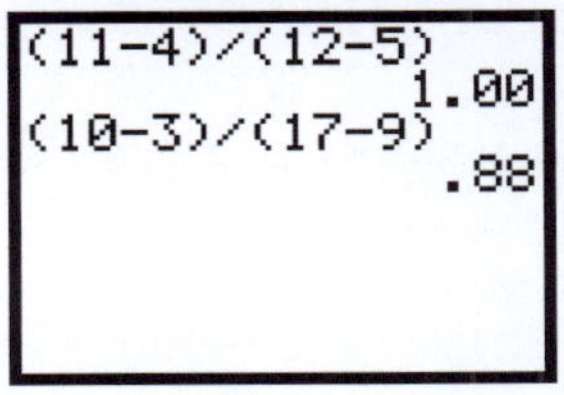

(6) Gradient $= \frac{(D - B)}{(C - A)}$.

(7) (a)

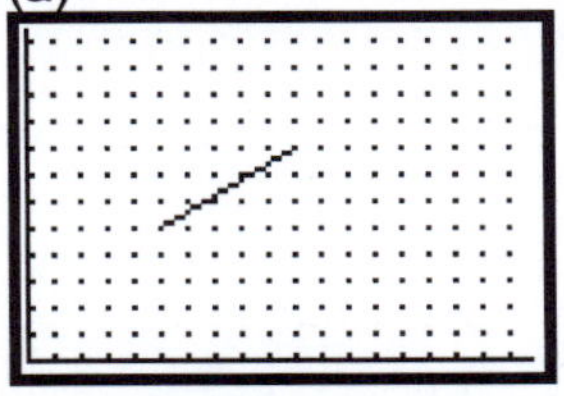

(b)

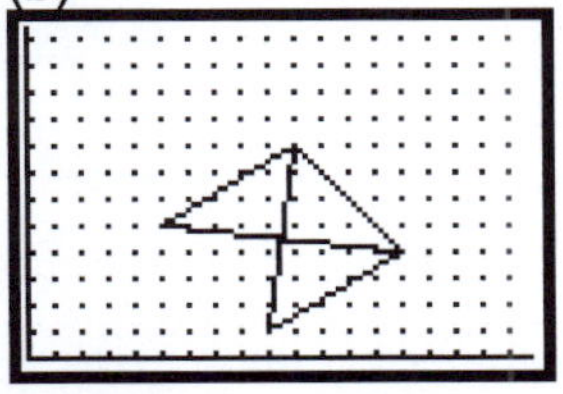

(c)-(d)

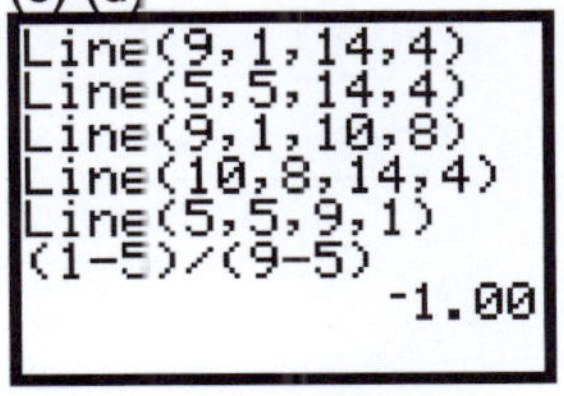

The important thing to notice is that the gradient of this line is negative.

(8) The points in part (b) give a negative gradient; those in parts (c) and (f) give a gradient of zero (the lines are horizontal) and those in parts (a), (d) and (e) give a positive gradient.

(9) Here are the key presses required to enter the program **GRADI**.

PROGRAM: GRADI	[PRGM] [◄] **1 GRADI** [ENTER]
:ClrDraw	[2nd] [DRAW] **1** [ENTER]
:Prompt A,B	[PRGM] [►] **2** [ALPHA] **A** [,] [ALPHA] **B** [ENTER]
:Prompt C,D	[PRGM] [►] **2** [ALPHA] **C** [,] [ALPHA] **D** [ENTER]
:Line(A,B,C,D)	[2nd] [DRAW] **2** [ALPHA] **A** [,] [ALPHA] **B** [,] [ALPHA] **C** [,] [ALPHA] **D** [)] [ENTER]
:Text(0,0,"GRADIENT=", (D–B)/(C–A))	[2nd] [DRAW] **0 0** [,] **0** [,] [2nd] [A-LOCK] **"GRADIENT** [2nd] [TEST] **1** [ALPHA] **"** [,] [(] [ALPHA] **D** [–] [ALPHA] **B** [)] [÷] [(] [ALPHA] **C** [–] [ALPHA] **A** [)] [)]

Finally, press [2nd] [QUIT] to return to the home screen.

(10) Here are the key presses required to change the program **GRADI** to its new form.

Select the program for editing	[PRGM] [►] and select **GRADI**
Remove unwanted commands	[▼] [CLEAR] [▼] [CLEAR]
Create four blank lines	[2nd] [INS] [ENTER] [ENTER] [▲] [▲] [▲]
:randInt(-4,4)→A	[MATH] [◄] **5 ⁻4** [,] **4** [)] [STO►] [ALPHA] **A** [ENTER]
:randInt(0,3)→B	[MATH] [◄] **5 0** [,] **3** [)] [STO►] [ALPHA] **B** [ENTER]
:randInt(-4,4)→C	[MATH] [◄] **5 ⁻4** [,] **4** [)] [STO►] [ALPHA] **C** [ENTER]
:randInt(-3,0)→D	[MATH] [◄] **5 ⁻3** [,] **0** [)] [STO►] [ALPHA] **D** [ENTER]
Insert **:Pause**	[▼] [2nd] [INS] [PRGM] **8**

Finally, press [2nd] [QUIT] to return to the home screen.

(11) (a) Suppose you choose the points ($^{-}2$, $^{-}1$) and (0,0.2).
The gradient is (0.2–$^{-}1$)/(0–$^{-}2$)=0.6. You always get 0.6, whichever points are chosen.
When you multiply out the brackets you get Y1= 0.6X+0.2.
The gradient of the line is equal to the coeficient of X.

(b) Multiplying out the brackets and comparing the gradients in each case gives:
Y2=0.5X–0.25, gradient 0.5, Y3=0.67X+2, gradient 0.67 (or 2/3)
Y4=$^{-}$4X+4, gradient $^{-}$4 Y5=2 (or 0X+2), gradient 0
Y6=$^{-}$2X+2, gradient $^{-}$2.

(c) The gradient of the straight line with equation Y=mX+c is m
(and c is the intercept on the Y axis).

So finally ...

◊ If two lines have gradients of 2 and $^{-}$2 they are equally steep, but one is going up and the other going down as you move from right to left. The point of this question is to bring out the distinction between the steepness (always a positive quantity) and the gradient (can be positive or negative). There's a similar distinction between speed and velocity.

◊ A line joining the origin to the point (S,T), has a gradient of $\frac{T-0}{S-0}=\frac{T}{S}$.

◊ The gradients of all horizontal lines are zero because

$$\frac{\textit{Distance up}}{\textit{Distance along}}=\frac{0}{\textit{Distance along}}=0$$

The gradients of all vertical lines are undefined because

$$\frac{\textit{Distance up}}{\textit{Distance along}}=\frac{\textit{Distance up}}{0}$$ and division by zero is impossible.

You might have said that the gradient is infinity, and this answer is OK too (as is negative infinity!)

Unit 2 A close look at curves

(1) Notice that the power of 3 can be entered using option 3 in the Math menu: press [MATH] **3**. An alternative is to use [^] **3**. Similarly the power of 2 can be entered using [x^2] or [^] **2**.

Resetting the defaults should ensure that your graph looks like the one shown in the text. If you don't do this you may have grid points still showing from Unit 1. They are not needed in this unit.

(2) The term pixel is used to describe the tiny individual dots that make up the calculator's screen. The values produced by Y1(-4.7) etc. are the same as the Y coordinates shown when you use [TRACE].

Brain stretcher When operating in the default mode, **Connected**, the calculator draws a straight line connecting each of the plotted points. Changing this setting to **Dot** shows only the points whose coordinates have been calculated. Changing **Xres** to a value such as 5 forces the calculator only to evaluate Y1 values for every 5th pixel along the X axis.

(3) It is well worth getting to grips with this procedure, as it will be used again in later units. Using a scatterplot rather than a linegraph means that the points are all plotted but they are not connected together. (Compare the use of **Connected** and **Dot** – see the Brain stretcher above.)

(4) (a) The curve is not as steep at X=1.5 as it is at X=0.

(b) The gradient of the curve is zero at roughly X=$^{-}$1.7 and X=2.1.

(c) The gradient is positive between about X=$^{-}$1.7 and X=2.1.

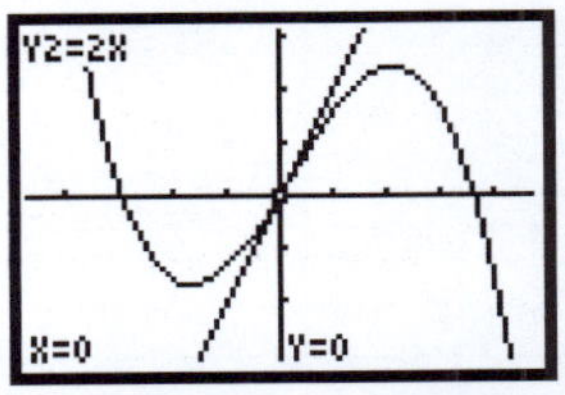

(d) The gradient is negative for values of X less than about X=$^{-}$1.7 and more than about X=2.1.

(e) The gradient of Y1 appears to be less than that of Y2=2X at X=0.

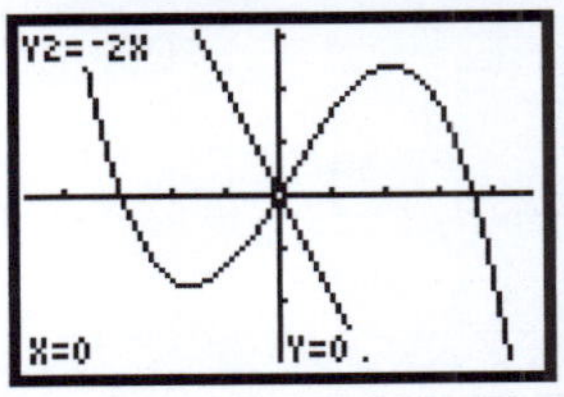

(f) Drawing the line Y2=$^{-}$2X gives some assistance. It seems that the curve might have the same gradient at about X=$^{-}$2.5 and about X=2.5.

(g) The curve seems to be steepest as it leaves the screen at about X=-3.7 and about X=4.2.

(h) The answer is not the same as for part (g)! The points at the steepest part of the curve have large *negative* gradients. The *maximum* gradient probably occurs at about X=0.

(5) (a) The highest value of Y appears to be Y=2.4122214.
(b) The lowest value of Y appears to be Y=$^{-}$1.694526.

(6) (a) You may get a display looking something like this.

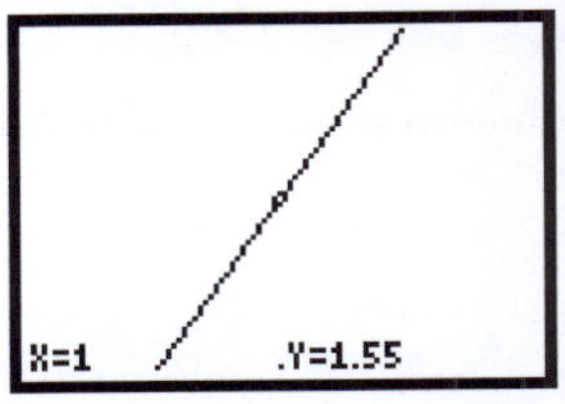

(b) Whichever point on the curve you choose, zooming in produces a straight line.

(7) (a) X=2 (b) X=$^{-}$2 (c) X=0 (d) X=$^{-}$3 (e) X=$^{-}$1 (f) X=4 (g) X=3 (h) X=1.

(8) The graphs and comments about the gradients are on the next page. In most cases, zooming in on the curve would make it resemble a straight line; the only exceptions are the points where the curve does not exist (the so-called discontinuities) in parts (a), (b) and (i).

(a)

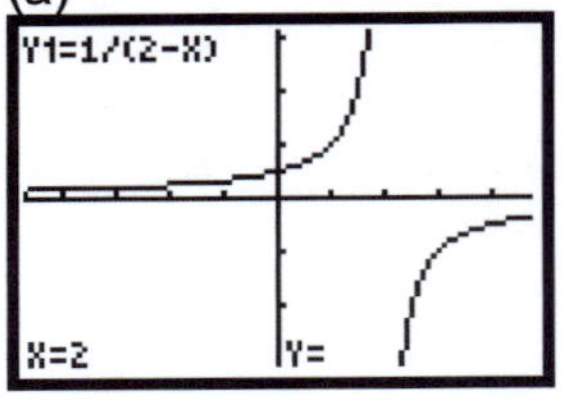

The gradient is always +ve, except at X=2 where the curve has no real value.

(b)

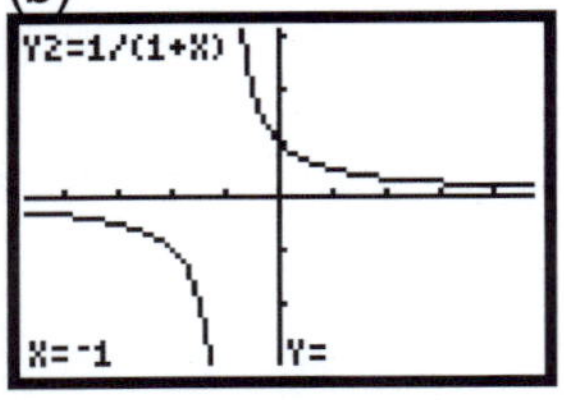

The gradient is always –ve, except at X=⁻1 where the curve has no real value.

(c

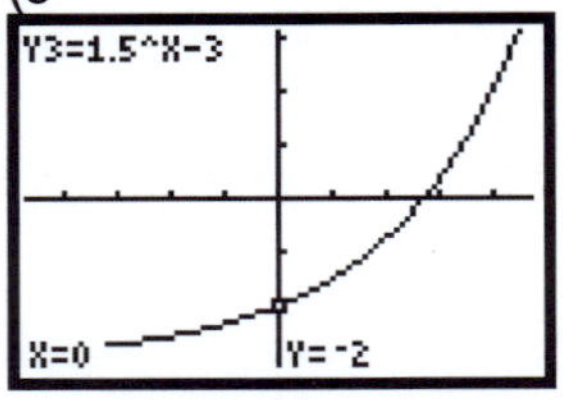

The gradient is always +ve and increases for increasing values of X.

(d)

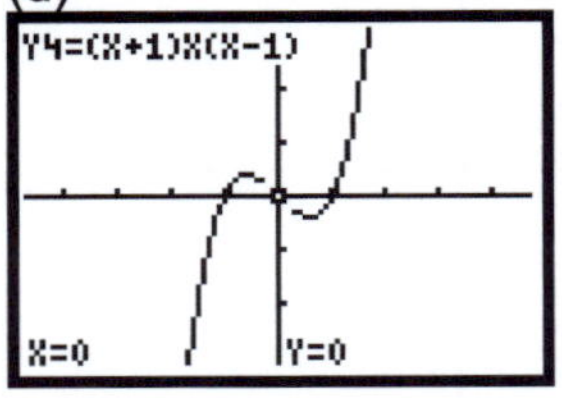

The gradient has both +ve and –ve values with two points where it is zero.

(e)

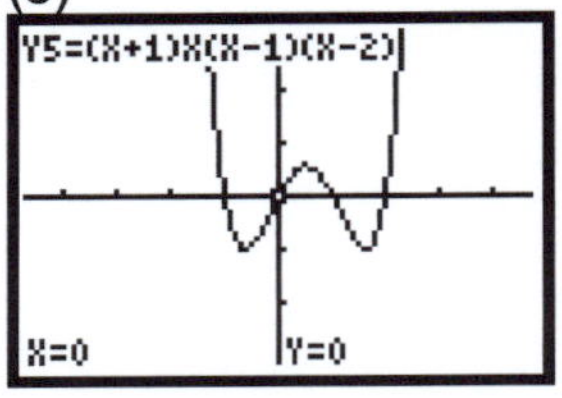

The gradient has both +ve and –ve values with three points where it is zero.

(f)

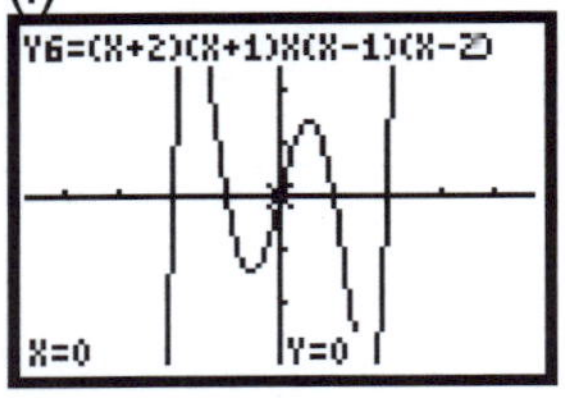

The gradient has both +ve and –ve values with four points where it is zero (two of them off the screen).

(g)

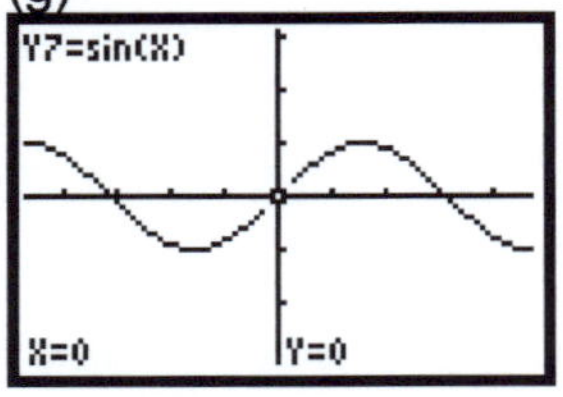

The gradient has both +ve and –ve values with lots of points (off the screen) where it is zero.

(h)

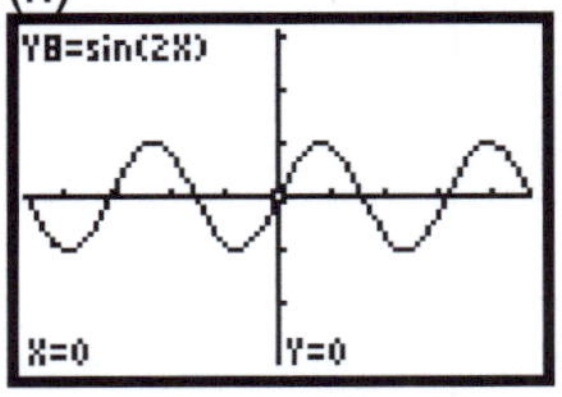

The gradient has both +ve and –ve values with lots of points (off the screen) where it is zero.

(i)

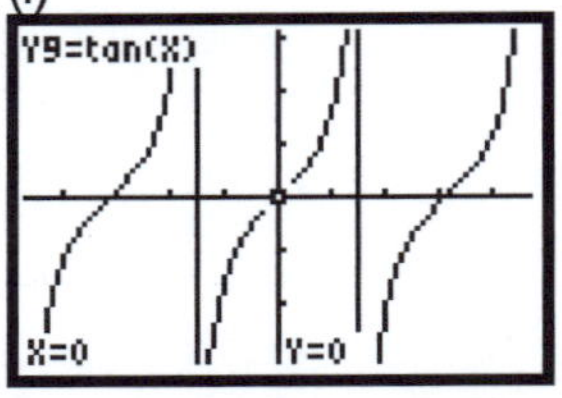

The gradient is always +ve.

Here the display is rather misleading – the vertical lines at about ±1.5 should not be drawn at all. Try zooming in and you find that there are no values of tan(X) there.

(j)

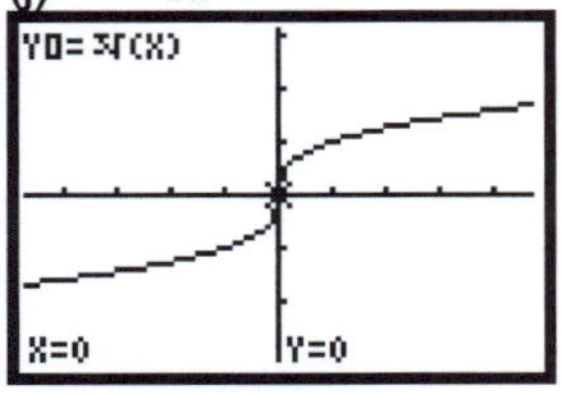

The gradient is always +ve.

So finally ...

◊ In order to draw the graph of a function the calculator has to know the y-coordinates corresponding to the x-coordinates of every pixel across the screen (assuming **Xres** has a value of 1). These are calculated using the functions selected on the Y= screen and the points plotted. If Connected has been selected in the MODE menu, short straight lines are drawn to join up the points.

◊ It is possible to have points with gradients of zero that are not "hills" or "valleys". For example, a curve can resemble a horizontal ledge on a cliff face. For example, try drawing the graph of **Y1=(X – 2)^3+1** Such points are known as 'points of inflexion' and they are the subject of Unit 9.

◊ In a sense it is both true and false that a curve is made up of lots of tiny straight lines. Certainly curves drawn on the calculator's screen are formed from lots of tiny straight lines; this is not a true curve – it only has the appearance of being one. A true mathematical curve consists of a continuous series of points that have no width and are no distance apart, but this situation can only exist is one's imagination!

◊ The calculator command **ΔX** is option 8 in the VARS Window menu. It gives the current horizontal distance between pixels on the graphing screen, and this of course is determined by the current window settings. So after using **ZDecimal**, ΔX is 0.1, but if you zoom in it gets progressively smaller.

◊ Zooming in on a curve does not really produce a straight line, only the appearance of one. The closer you look at the curve the smaller the difference between the coordinates of adjacent points, and the more the gradient seems to be constant.

Unit 3 Gradient of a curve at a point

(1) (a) At X=1, Y1=1.55.

(b) At X=2, Y1=2.4.

(c)

```
(2.4-1.55)/(2-1)
                .85
```

(2) (e) You see the values on the right for the gradient as X gets closer to X=1:

```
     1.0315
      1.086
     1.1375
      1.186
     1.2315
      1.274
     1.3135
```

But when the two points are on top of each other, X–A=0 and you get the error message: DIVIDE BY 0.

(f) Looking at the graph you can see that the gradient is steeper as the second point gets closer to the first. This is confirmed by the increasing values calculated on the home screen.

(3) (a) As (X,Y) moves further away from (A,B) to the left, the gradient continues to increase.

```
(Y-B)/(X-A)
     1.3835
      1.414
     1.4415
      1.466
     1.4875
```

(b) When X=1.1 the gradient is 1.3135. When X=0.9 the gradient is 1.3835.

(4) (g) The next few estimates are shown here. They appear to be converging towards 1.35.

```
1.347806641
1.349452759
1.349863258
1.349965819
1.349991454
1.349997854
1.349999472
```

(5)

```
(Y-B)/(X-A)
        1.3835
    1.35865625
   1.352181641
   1.350546509
   1.350136696
   1.350034178
```

(6) (a) The valley point occurs closest to X=⁻1.7.

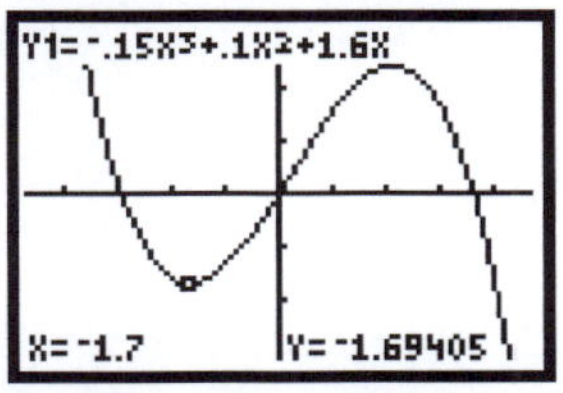

(b) Zooming in on the right of X=⁻1.7 gives the series of estimates shown here.

```
(Y-B)/(X-A)
         .0445
     -.01896875
   -.0350996094
   -.0391488037
   -.0401621322
   -.0404155284
```

(c) The hill-top point occurs closest to X=2.1.

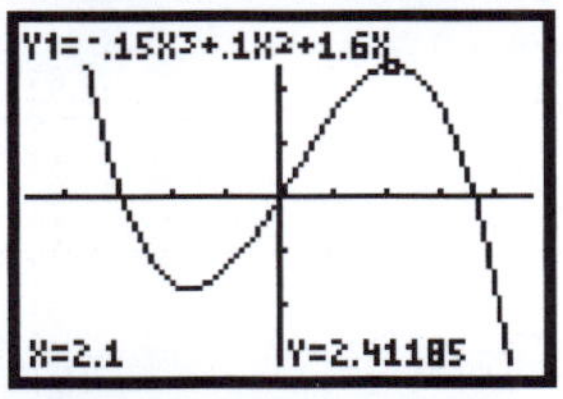

Estimates of the gradient at X=2.1 are shown here.

```
(Y-B)/(X-A)
         -.0505
       .01428125
    .0302128906
    .0341793213
     .035169899
    .0354174799
```

(7) (b) It seems likely that the gradient at the origin is actually 1.

```
X→A:Y→B
              0
(Y-B)/(X-A)
    .9971466573
    .9998215228
    .9999888446
    .9999993028
```

(c) The symmetry of the curve suggests that the gradients where it cuts the X axis are ⁻1 and 1.

This can be checked for example as shown here.

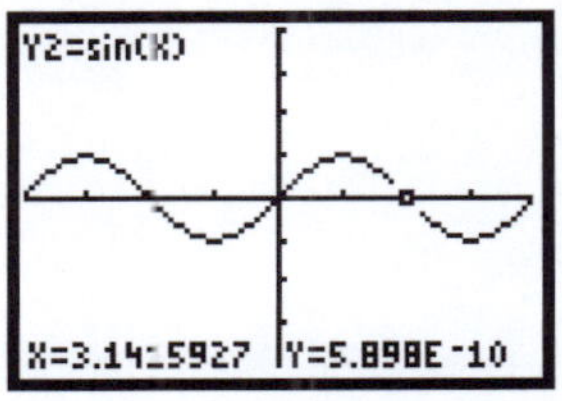


```
X→A:Y→B
      5.898E-10
(Y-B)/(X-A)
   -.9971466574
   -.9998215228
   -.9999888446
   -.9999993028
```

Brain stretcher

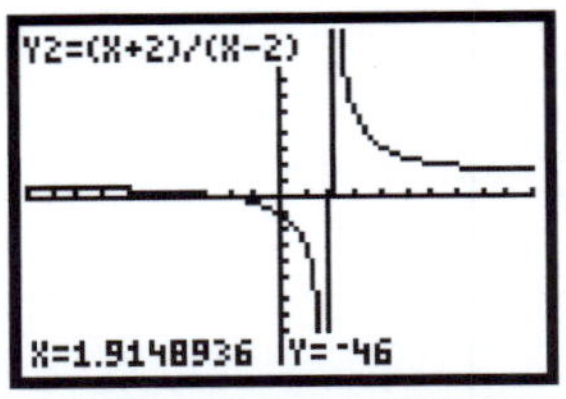

The **ZStandard** window setting produces a steep, nearly vertical, line at X=2, but this is caused by the calculator evaluating Y2 at points just below and just above X=2 and then joining those two points. In fact Y2 has no value at X=2. (Check using the Table or by pressing [TRACE] **2** [ENTER]). There is a so-called *discontinuity* at X=2. It follows that it cannot have a gradient at X=2.

```
X→A:Y→B
          -46.00000165
(Y-B)/(X-A)
           368.1666924
```

Using the method set out in the Brain stretcher makes it appear that the gradient is about 368 but this method also uses points on either side of the discontinuity.

So finally ...

With the functions used in this unit it has generally been possible to predict the gradient of curves after zooming in just a few times. This is because the gradients have had exact values such as 1.35. As you can see from the screen displays in Solutions 4 and 5, it quickly becomes obvious to what gradient value the series of estimates is converging.

However, with other functions it is by no means so easy. To what value do you think this series of estimates is converging? It looks as if it may 0.7785 (to 4 d.p.) but if greater accuracy is required more zooming will be necessary.

```
.7824653213
.7795199869
.7787842614
.7786003632
.7785543899
.7785429292
.7785401167
```

Unit 4 Tabulating and programming gradients

(1)

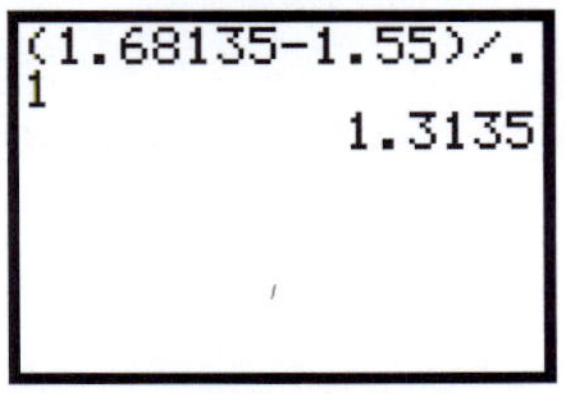

(3)(a)

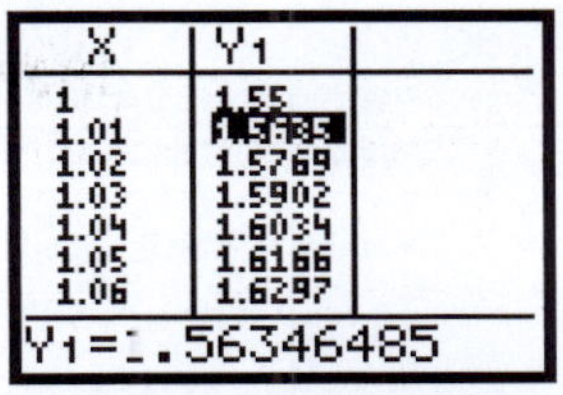

(b) Notice that the TBLSET menu displays values of **ΔTbl** in standard form for values .0001 and smaller.

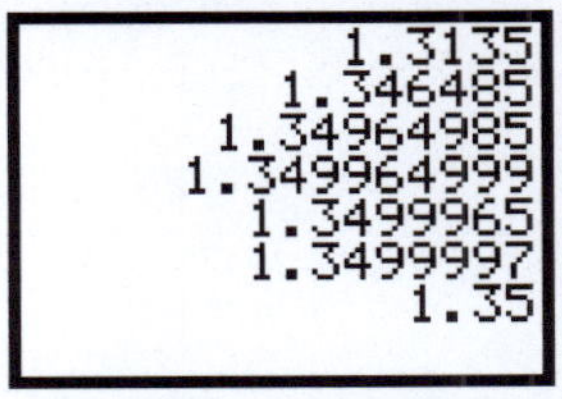

(4)

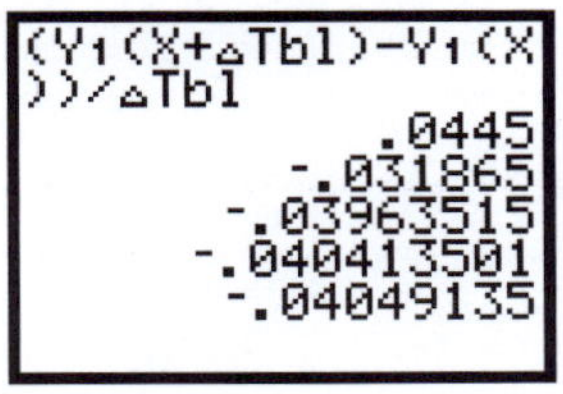

You may have wondered why the approximations for the gradient change from a positive to a negative sign. The reason is that X=⁻1.7 is very close to the bottom of the "valley point" and on the left of it. When **ΔTbl** = 0.1 the second point is to the right of the valley point and above the point of interest so there is a postive gradient. For smaller values of **ΔTbl** the second points are to the left of the valley point and below the point of interest so there is a negative gradient.

(5)

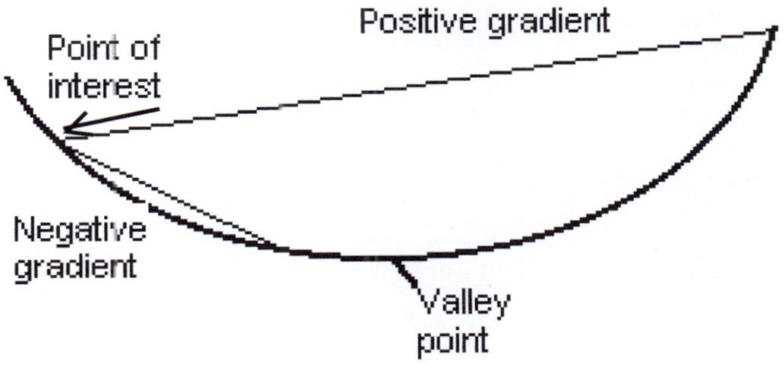

(6) (b) At X=3

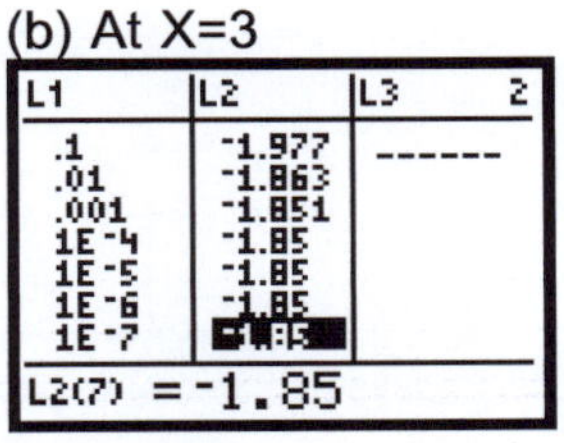

At X=4

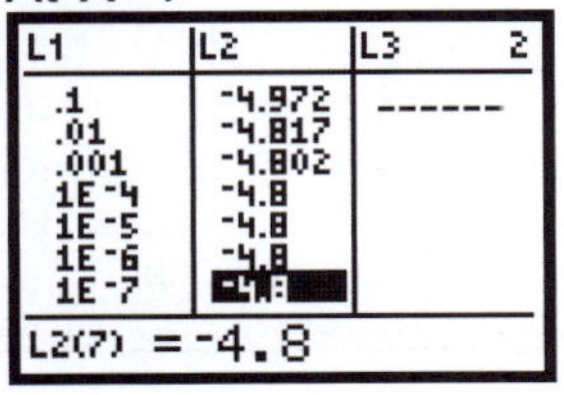

At X=5

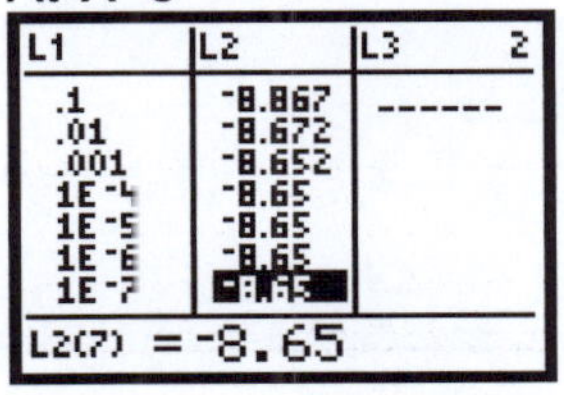

In each case, by the time **ΔTbl** = E⁻7 you can be pretty sure of the value of the gradient.

Brain stretcher

For values of **ΔTbl** smaller than E⁻10 the calculator begins to have problems evaluating the expression in the sixth line of the program.

For example, look at these values for the gradient at X=5.

This is because the calculator's limits of computational accuracy have been exceeded. (See TI-83 Plus Guidebook, page B-11).

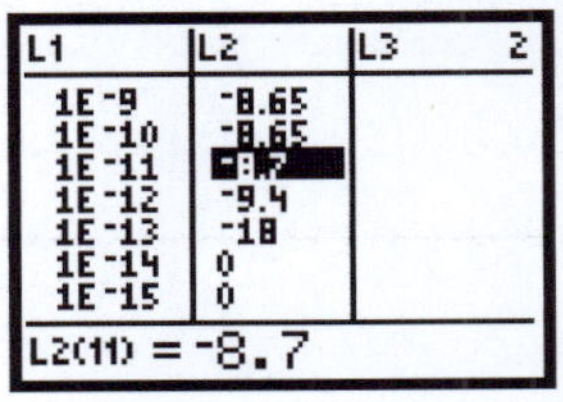

(8) (a) ⁻3.05 (b) ⁻0.6 (c) 0.95 (d) 1.60000001 (e) 0.2 (f) ⁻1.85 (g) ⁻4.8

So finally ...

In Unit 3 you repeatedly zoomed in on the graph and were able to use smaller and smaller values of ΔX (the pixel width) in the gradient calculation. In this unit you used smaller and smaller values of **ΔTbl** instead. Otherwise the two methods are essentially the same.

Unit 5 Dee why by dee ex

(1) With a function stored in Y1, the y-coordinate corresponding to X is Y1(X) and the y-coordinate corresponding to X+δx is Y1(X+δx). So the change in the y-coordinate, known as δy is Y1(X+δx)–Y1(X).

(3) (a) −3.05 (b) −0.6000002 (c) 0.94999985 (d) 1.5999999

(e) 0.19999985 (f) −1.85 (g) −4.8

It is interesting to compare these values with those calculated using program **GRADIENT** in Unit 4.

(4) It is called a *local* maximum to stress the fact that it is the maximum in that locality, with lower values on either side. There may be other higher values further away at different parts of the curve.

For the local maximum, the gradient closest to zero (shown here) occurs at X=2.1.

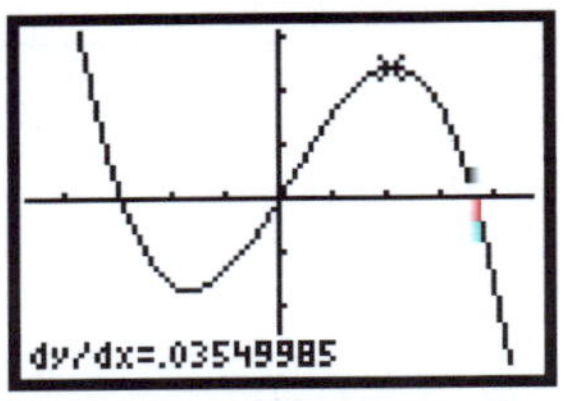

For the local minimum, the gradient closest to zero (shown here) occurs at X=−1.7.

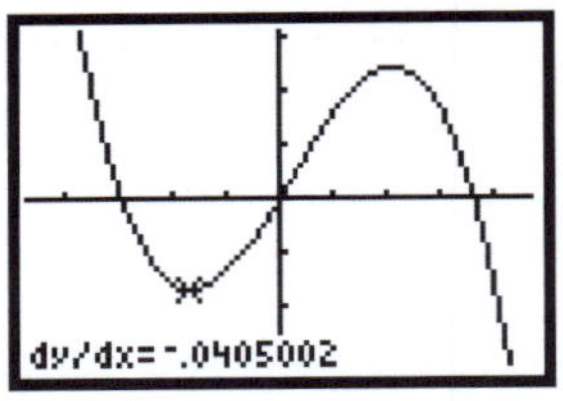

(5) The maximum gradient of this curve appears to be very close to 1 at X=0.

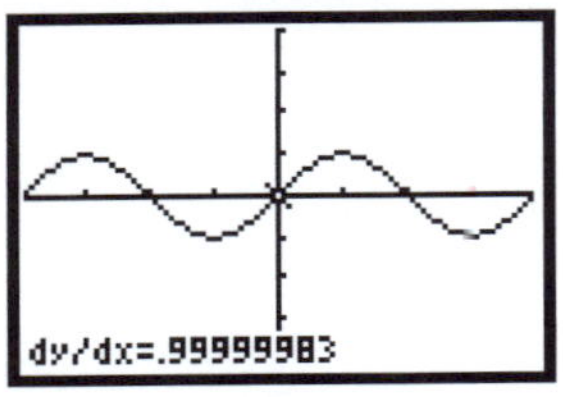

(6) With the **ZDecimal** window settings no value is plotted at X=2 and, as before, there is no calculated value for **dy/dx**.

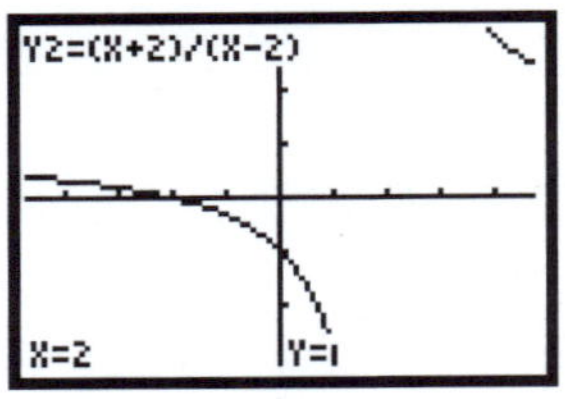

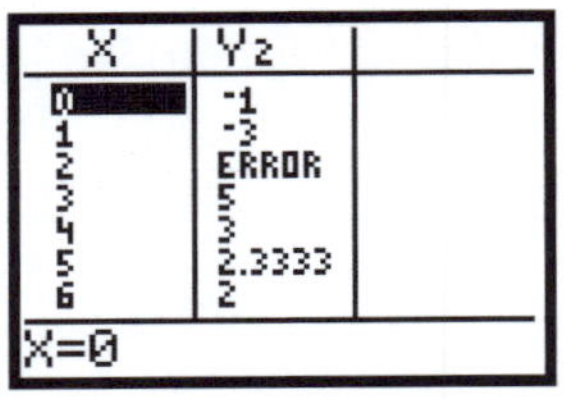

Brain stretcher

The curve on the left was produced by the function **Y2=10/((X+2)(X–2)(X–4))**, and the one on the right by the function **Y2=tan(X)**.

(8) You may have wondered why the gradient in the equation of the tangent is −3.05000015 whereas the gradient produced by **dy/dx** in Activity 3(a) was −3.05. You can see what is going on by doing the **dy/dx** calculation again and then going to the home screen and entering **Ans**. Clearly the calculator rounds the gradient when it displays **dy/dx** but, unaccountably, does not round it when displaying the equation of the tangent.

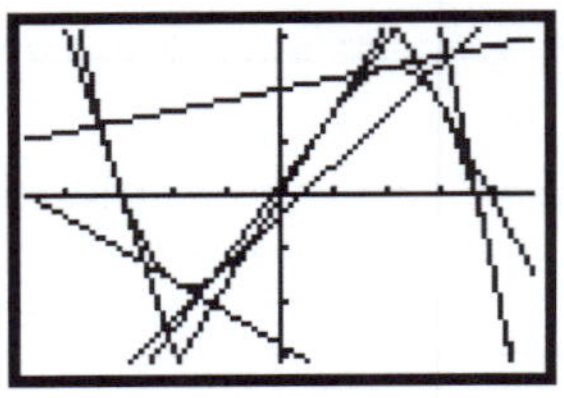

(10) (a) **dy/dx** =−0.7125 (to 4 d.p.)

Y=2.28125.

(b)

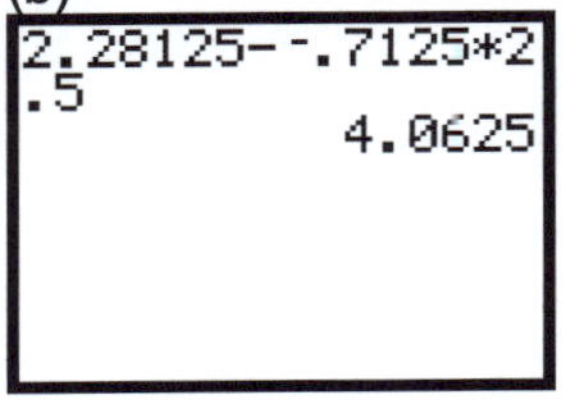

Alternatively, following the calculation of **dy/dx** you could enter the calculation like this.

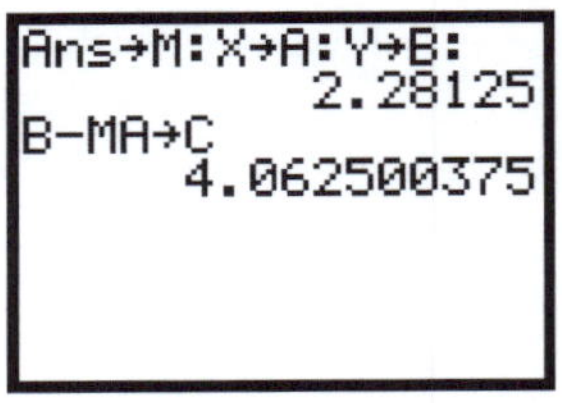

(c)

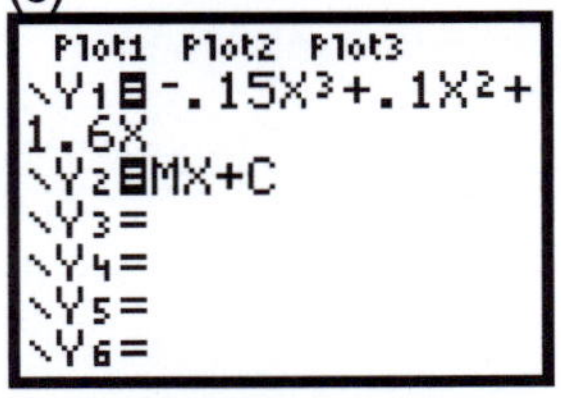

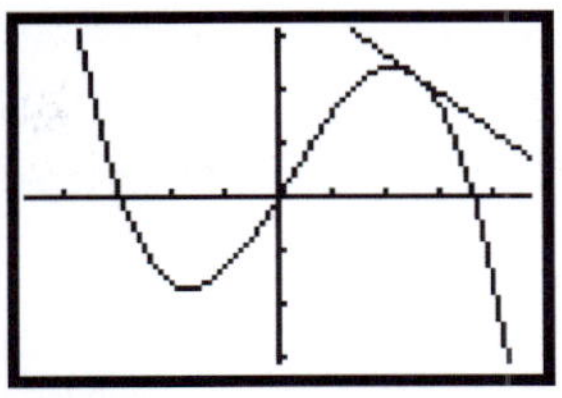

(d)

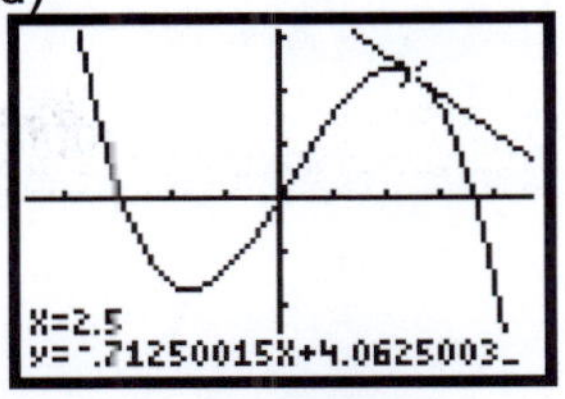

(11) Here are the key presses required to enter the program **LOTATANS**.

PROGRAM: LOTATANS	[PRGM] [◄] **1 LOTATANS** [ENTER]
:For(C,XMin,Xmax,5ΔX)	[PRGM] **4** [ALPHA] **C** [,] [VARS] **1 1** [,] [VARS] **1 2** [,] **5** [VARS] **1 8** [)] [ENTER]
:Tangent(Y2,C)	[2nd] [DRAW] **5** [VARS] [►] **1 2** [,] [ALPHA] **C** [)] [ENTER]
:End	[PRGM] **7**

Now press [2nd] [QUIT] to return to the home screen.

So finally ...

All the terms are explained in the unit so you should be able to check your understanding there.

Unit 6 nDeriv - how does it do it?

(2) (a)

At X=	2.0	2.2	2.4	2.6	2.8	3.0
nDeriv=	0.19999985	⁻0.13800015	⁻0.51200015	⁻0.92200015	⁻1.36800015	⁻1.8500015

(b)

At X=	5	50	500
nDeriv=	9	99	999

(3) (a)(b)(c)

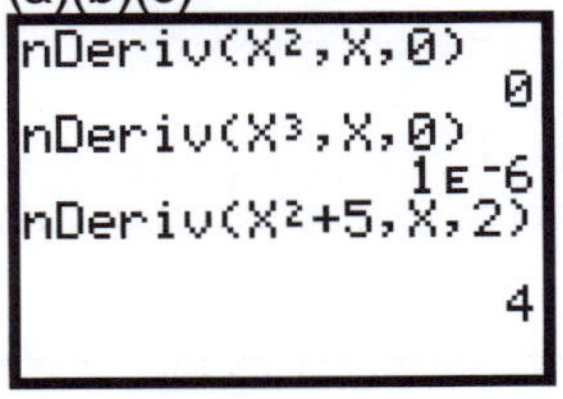

(d)(e)

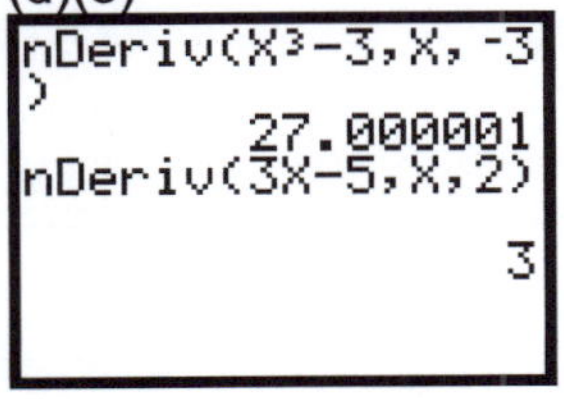

(f)(g)

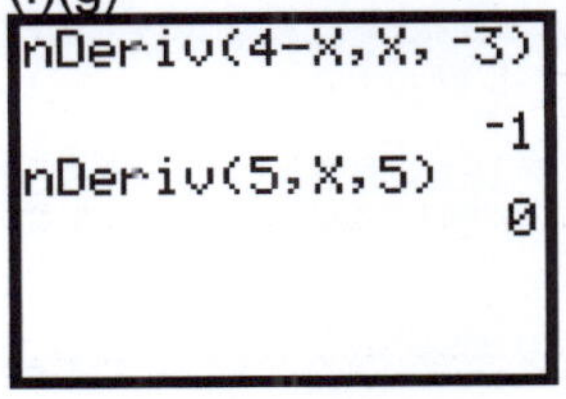

(h)(i)(j)

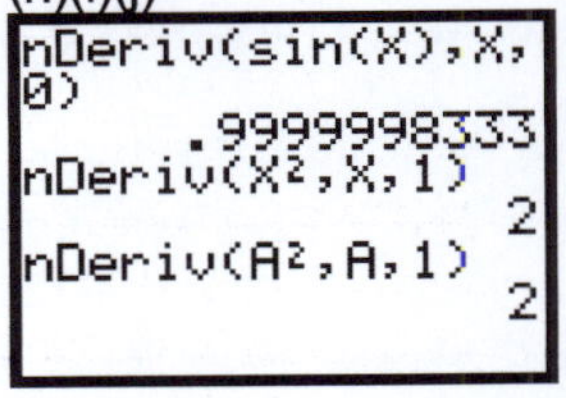

(4) (b)

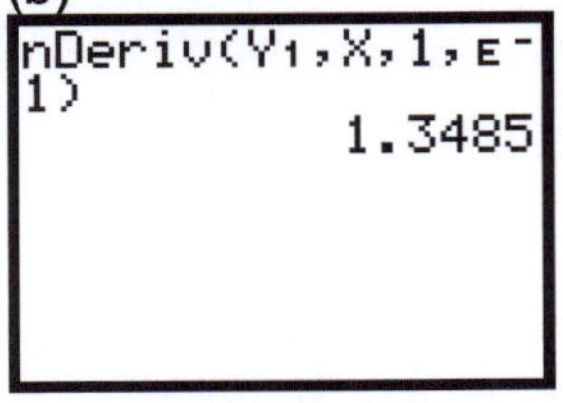

(c)

An alternative way of storing successive values is shown here.

The table below gives the values of **nDeriv** for decreasing incremental values. Notice that the second and subsequent values of **nDeriv** appear in list L3 on the list screen rounded to 1.35. Their more accurate values may be seen at the bottom of the screen when the cursor is moved over any particular value.

Increment	E⁻1	E⁻2	E⁻3	E⁻4	E⁻5	E⁻6
nDeriv	1.3485	1.349985	1.34999985	1.349999999	1.35	1.35

The same value of **nDeriv**, 1.35, is produced for smaller increments as far as E⁻12. However, after E⁻12, for even smaller increments **nDeriv** produces a value of 0.

(d) Recall that, when the default increment was used, **nDeriv(Y1,X,1)** gave a value of 1.34999985. This is the same value as was produced by **nDeriv(Y1,X,1, E-3)** so the default value must be E¯3 or 0.001. (This is the value stated in the calculator Guidebook page 2-7.)

(e) Larger increments produce less accurate approximations, for example **nDeriv(Y1,X,1,1)** produces a value of 1.2.

Brain stretcher

The traditional method produces an underestimate whenever the slope is decreasing to the right i.e. the graph is curving downwards (as shown in the diagram in the text), It produces an overestimate whenever the slope is increasing to the right, i.e. the graph is curving upwards. A perfect result is produced only for straight-line graphs. An erroneous estimate can be produced if there is a discontinuity very close to the point of interest.
The symmetric difference quotient method can produce both overestimates and underestimates for a variety of curves. A perfect result is produced for straight-line graphs and many other curves. An erroneous estimate can be produced if there is a discontinuity less than one increment to the right or left of the point of interest.

So finally ...

◊ **dy/dx** only produces a result on the graphing screen so it is useful when you have the graph in front of you and want to know the gradient at a particular point. **nDeriv** cannot be used on the graphing screen. It is useful on the home screen if you want to calculate the gradients of lots of points (as, for example, in Activity 2). It also allows you to calculate a more accurate gradient if necessary, by entering a smaller incremental value as an optional fourth argument. As you will soon see in Unit 7, **nDeriv** can also be used on the Y= screen in a very powerful way.

◊ We hope that you found it interesting to learn how the calculator has been programmed to calculate gradients – but this was not the main reason for including it in this book. Rather, working through the second part of the unit has hopefully given you extra understanding of gradients themselves, by focusing your attention on various alternative ways of estimating their values.

Unit 7 Gradient functions

(1) **nDeriv** has been calculated at the current particular value of X, and this varies as you move around the graphing screen. You can check what the current value is by entering X on the home screen. Your X value, and so your **nDeriv** value may differ from those shown here.

(3)

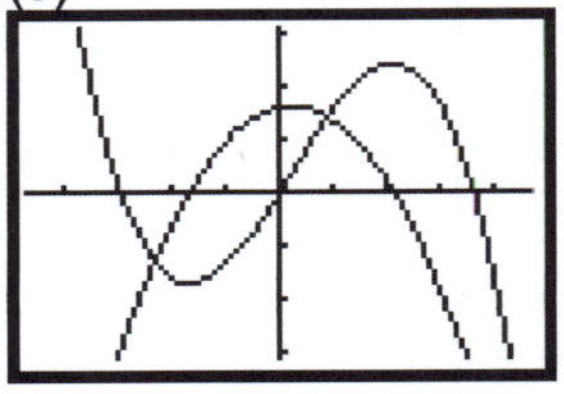

(4) You should see that the graphs of Y1 and Y2 are related in each of the ways described in the activity.

(5) The graph of the derived function has a local maximum (at about X=0.2) and is symmetrical about that point cutting the x-axis at about 2.1 and ¯1.7. This graph is typical of quadratic functions. Its actual equation will be calculated in the next few activities.

(8)

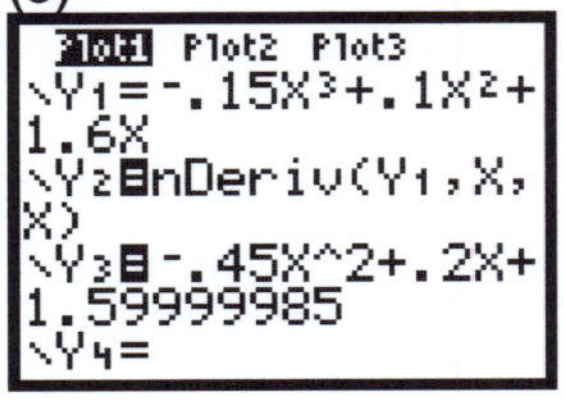

(9) (a)

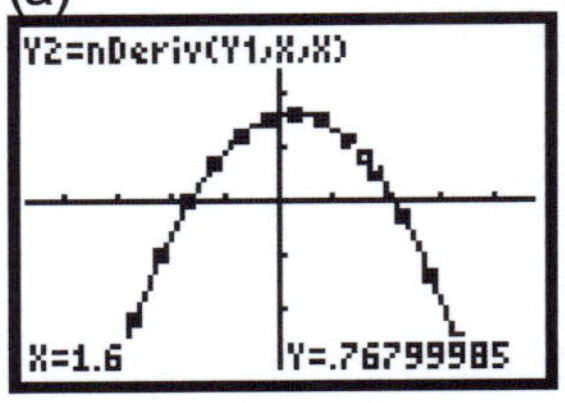

You can trace to any point on Y2, and then move to Y3 by pressing ▼. You will find no significant differences between the coordinates of the points on the two lines.

(b)

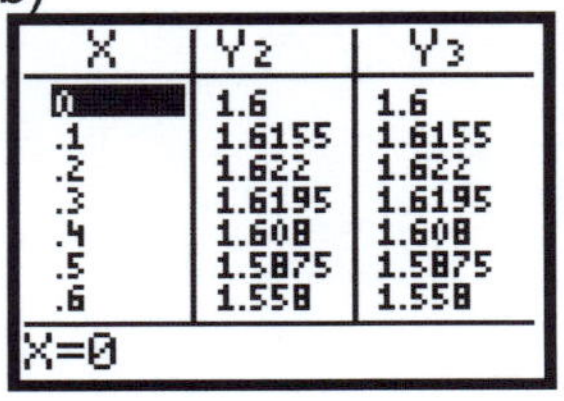

X	Y2	Y3
0	1.6	1.6
.1	1.6155	1.6155
.2	1.622	1.622
.3	1.6195	1.6195
.4	1.608	1.608
.5	1.5875	1.5875
.6	1.558	1.558

X=0

The tabulated values of Y2 and Y3 also show no significant discrepancies.

(10) The following activities will enable you to check any conjectures you have at this stage.

(12) (a) Whatever number replaces 1.6 as the coefficient of X in Y1 will be produced as the value of c in the quadratic regression equation. This is true even if a negative number is used. If you use very large numbers the graphs may be no longer visible on the ZDecimal viewing window. They could be viewed by changing the window settings but the above result still holds anyway.

(b) The coefficient of x in the derived function is twice the coefficient of x^2 in the original function. Notice that in order to move towards a general statement it is necessary to state the rule clearly in words. "It's always twice" may be what you think first, but you need to state clearly what "it" is, and also what is being doubled.

(c) The coefficient of x^2 in the derived function is three times the coefficient of x^3 in the original function. In symbols, if p is the coefficient of x^3 in the original function then the coefficient of x^2 in the derived function is 3p.

(d) An additional constant term has no effect at all on the derived function. The addition of any constant term simply moves the graph in a vertical direction and makes no difference to the shape of the curve; that is, it does not change its gradient. Therefore the gradient function is unchanged.

(e) If the original function were $y = px^3 + qx^2 + rx + s$

the equation of the derived function would be $\frac{dy}{dx} = 3px^2 + 2qx + r$

So finally ...

There were lots of important ideas in this unit, so it is well worth looking back to make sure you understand them all now.

Unit 8 Differentiation of polynomials

(1) (a) In this case $p=0$, $q=0.7$, $r=2.2$ and $s={}^{-}1$, so $\frac{dy}{dx} = 3px^2 + 2qx + r = 1.4x + 2.2$.

(b)

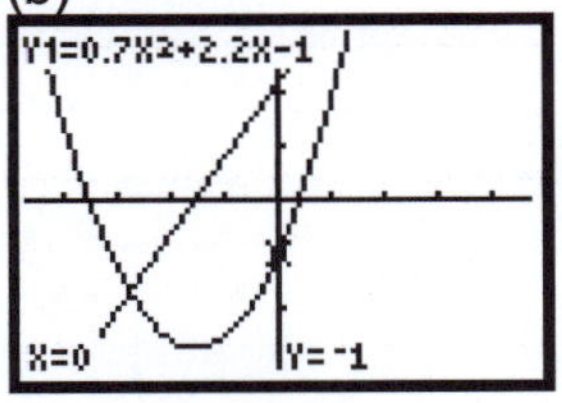

As expected, Y1 has the distinctive shape of a quadratic function.

This confirms the answer to part (a).

(c)

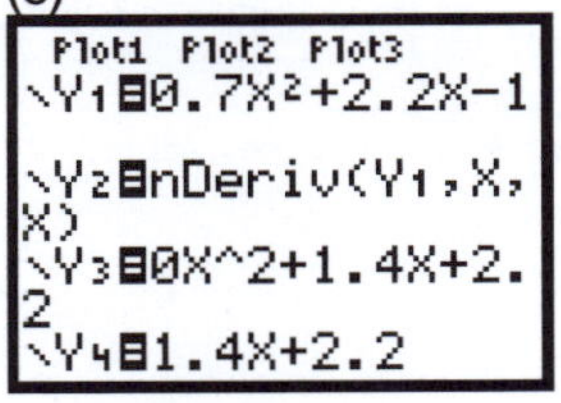

The **LinReg** command has produced the same result as in parts (a) and (b).

(2) (a) In this case p=0, q=0, r=0.3 and s=1,

so $\frac{dy}{dx} = 3px^2 + 2qx + r$

$= 0.3.$

(b)

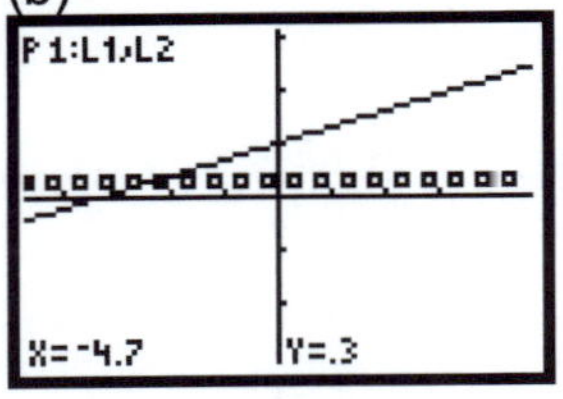

Program **SCATTER** produces points in a horizontal line. You can see at a glance that this time the derived function is a straight line with equation $\frac{dy}{dx} = 0.3$.

(c)

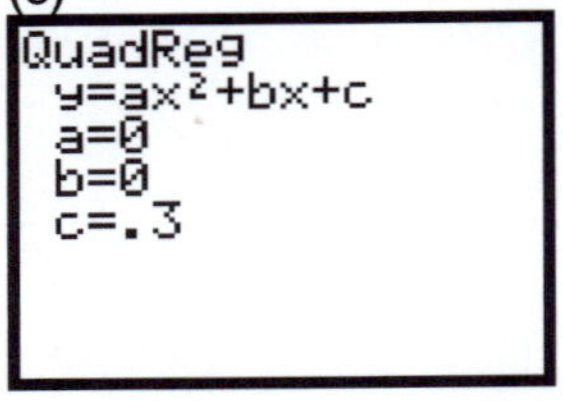

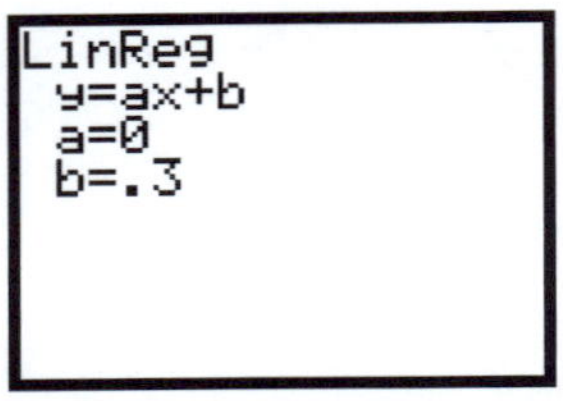

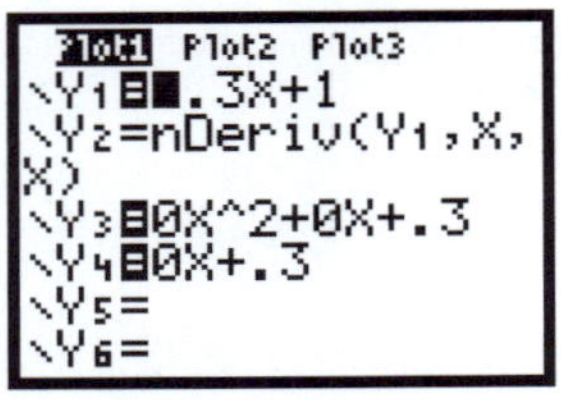

Either **QuadReg** or **LinReg** can be used as shown here.

(3) (a) Extending the pattern suggests that $\frac{dy}{dx} = 4ax^3 + 3px^2 + 2qx + r$.

(b) This suggests that if $y = 0.1x^4$ then $\frac{dy}{dx} = 0.4x^3$.

(c)

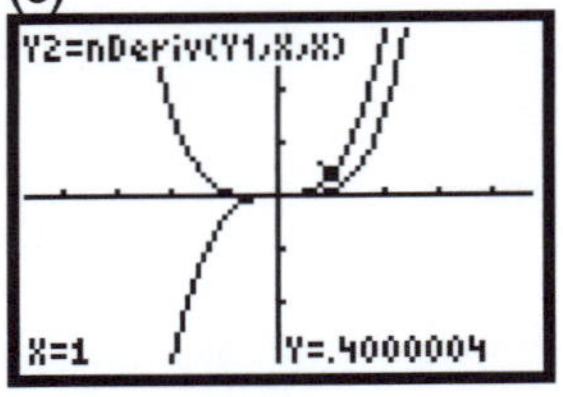

The graph of the derived function certainly has the shape of a cubic function such as y=0.4x^3 and the point traced here fits that function.

(d)

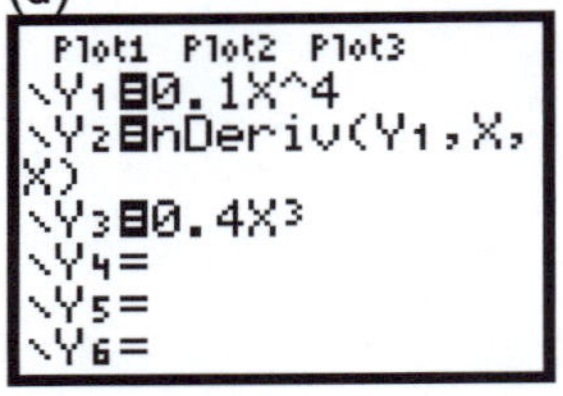

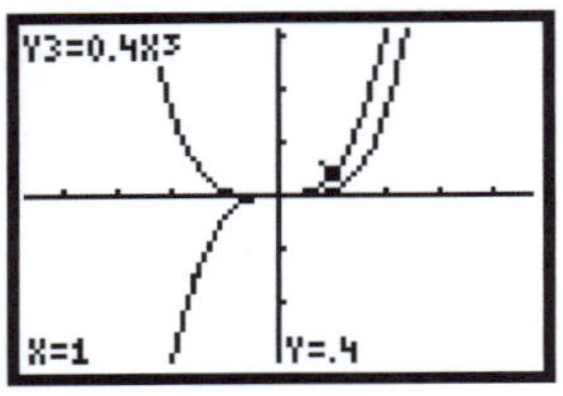

Y2 and Y3 have the same graphs.

(e)

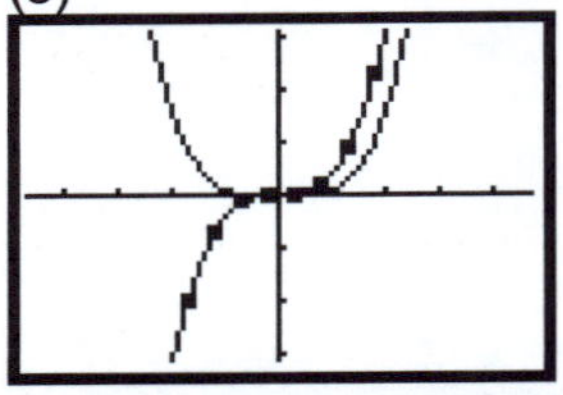

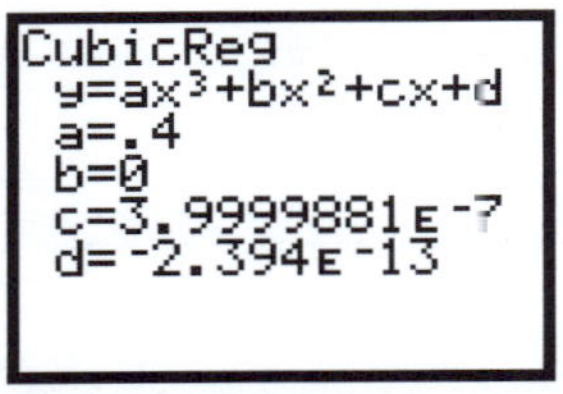

The values of **c** and **d** are very close indeed to zero, so this provides further confirmation that if $y = 0.1x^4$ then $\frac{dy}{dx} = 0.4x^3$.

(4)

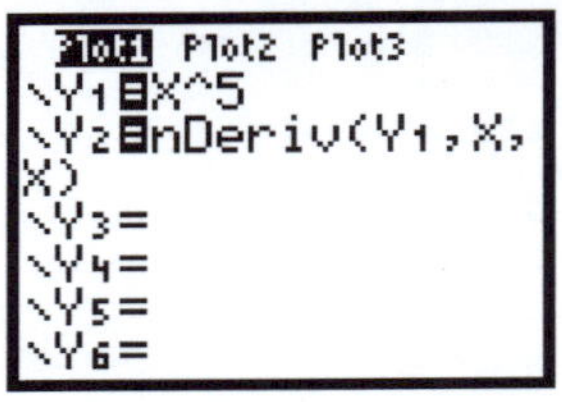

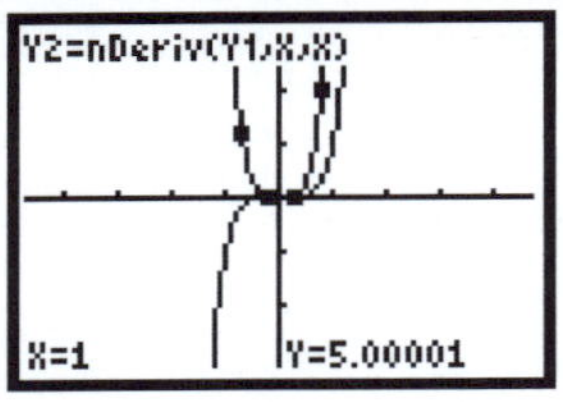

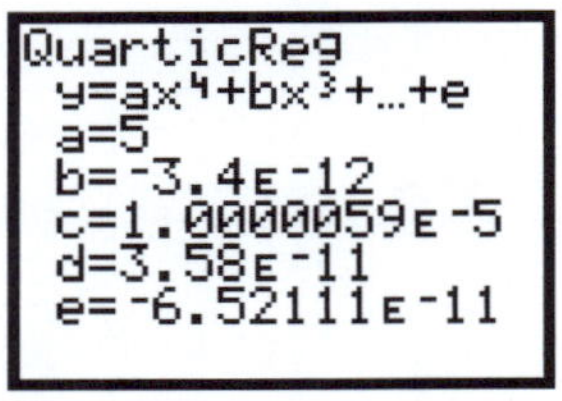

Since b, c, d and e are very close to zero, this suggests that if $y = x^5$ then $\frac{dy}{dx} = 5x^4$.

(5)

$y = ax$	$y = ax^2$	$y = ax^3$	$y = ax^4$	$y = ax^5$	$y = ax^6$
$\frac{dy}{dx} = a$	$\frac{dy}{dx} = 2ax$	$\frac{dy}{dx} = 3ax^2$	$\frac{dy}{dx} = 4ax^3$	$\frac{dy}{dx} = 5ax^4$	$\frac{dy}{dx} = 6ax^5$

The general rule looks as if it will be: if $y = ax^n$ then $\frac{dy}{dx} = nax^{n-1}$.

Comparing Y2 and Y3 can be done either using their graphs or their tabulated values as shown.

(a)

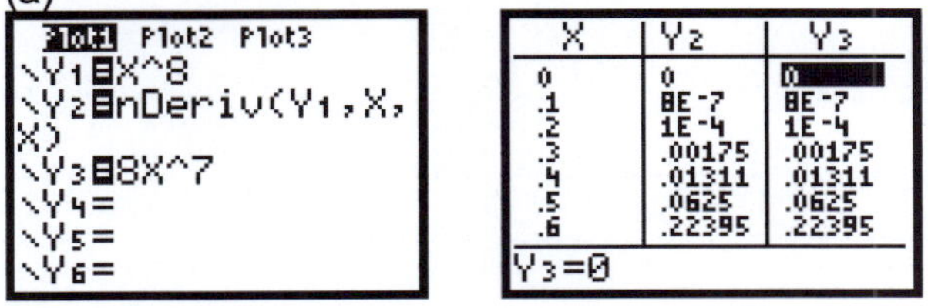

(b)

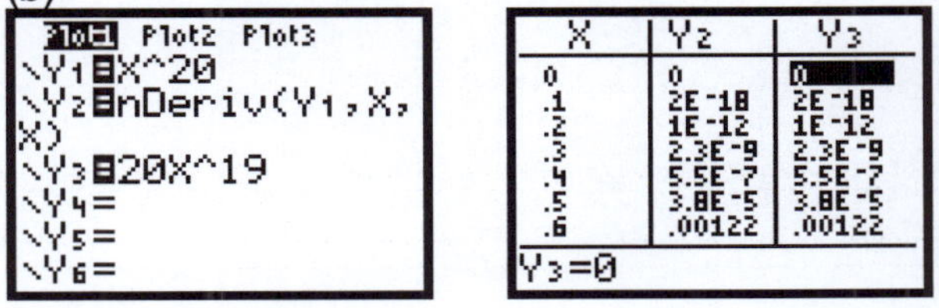

(c)

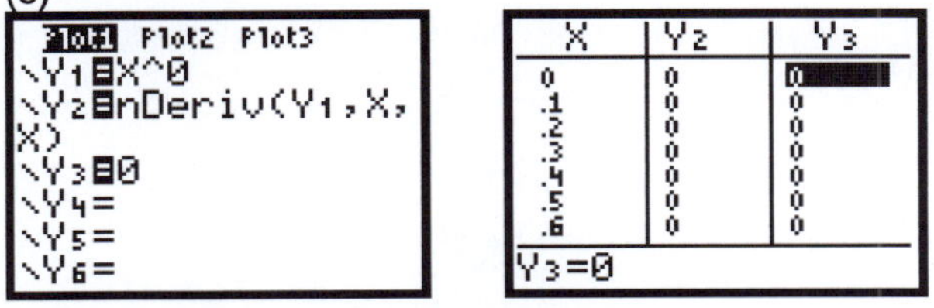

(d)

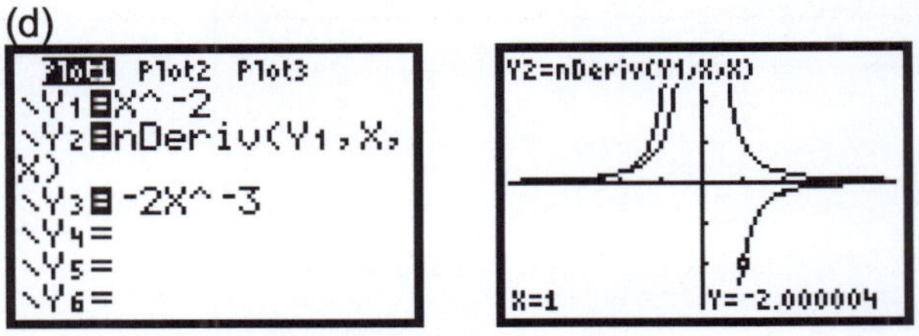

Notice that in part (d) there is a discontinuity at X=0, and hence no value for the gradient there.

(6) In each case is is only necessary to store the new value in N and then look at the tabulated values of Y2 and Y3.

(a)

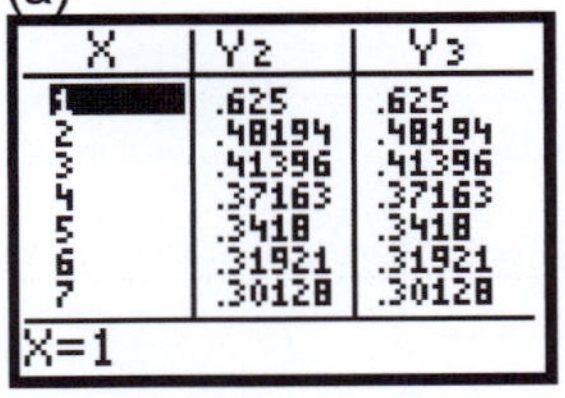

(b)

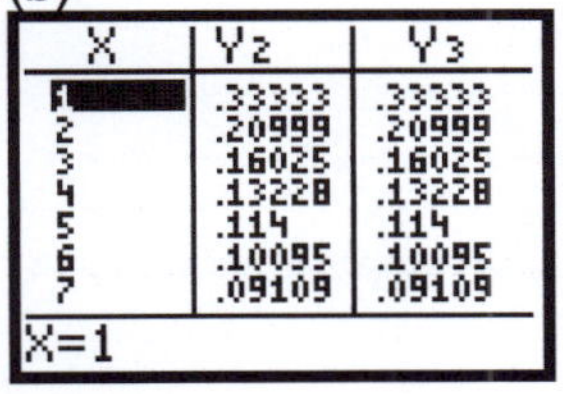

(c)

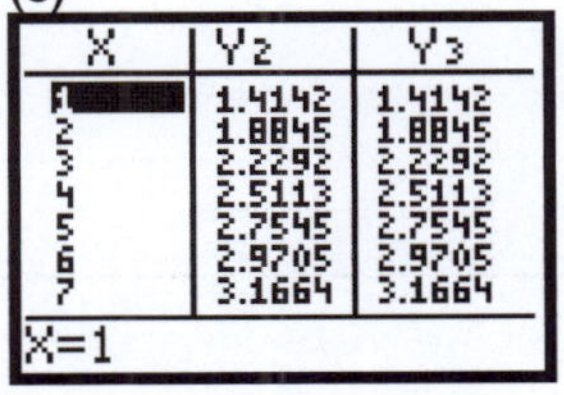

(d)

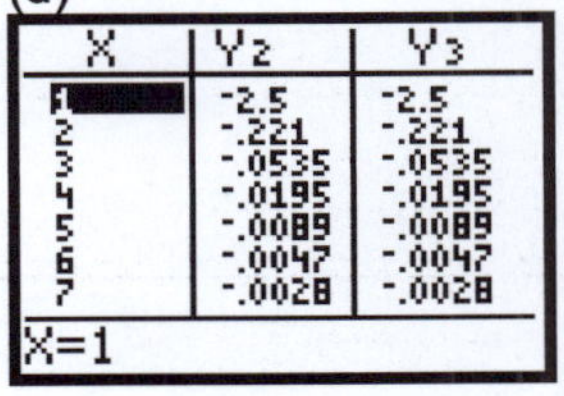

You may have noticed that, in Activity 6 parts (c) and (d), errors occur if you look at values of Y2 and Y3 corresponding to X≤ 0. The reason is that the original function does not exist for these values of X – check by looking at the graphs of Y1 in these cases.

So finally ... We hope that you are convinced about the general rule for differentiating functions that are powers of x, but it is important to point out that it has not been *proved* to be true for all values of n. All that you have done in this unit is show, using the calculator, that it is true for every value of n that you have happened to try. A formal proof is certainly possible and most traditional calculus textbooks will include it.

Notice also that it is possible to prove that if you have a sum of separate terms forming a polynomial function, it is valid to differentiate the function, term by term, simply taking account of the plus or minus sign between the terms. Again this hasn't been proved in this book, but we hope you have seen enough examples to feel comfortable with this rule too.

Unit 9 Points of inflexion

(1) (a)

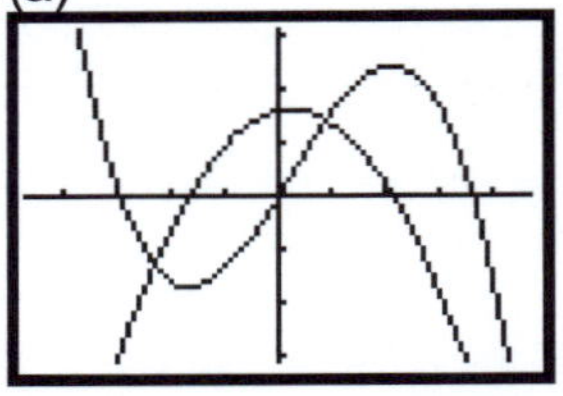

(b)

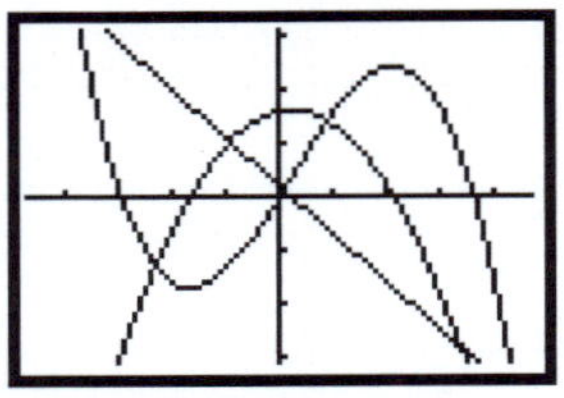

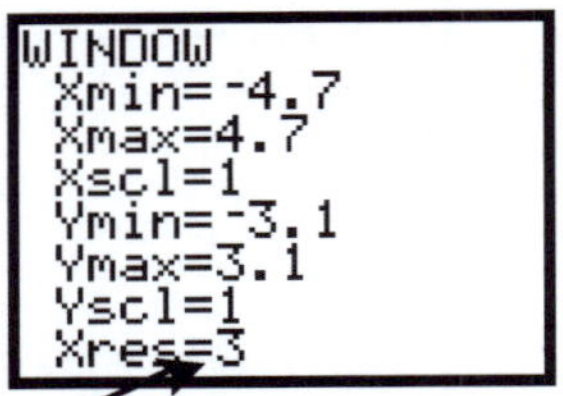

If you want to speed the process up you can change **Xres** from 1 to a larger number (see the Brain Stretcher in Unit 2). Of course, the curve becomes a little less smooth as a result.

(2) (a) Y3 appears to be close to zero at X=0.2.

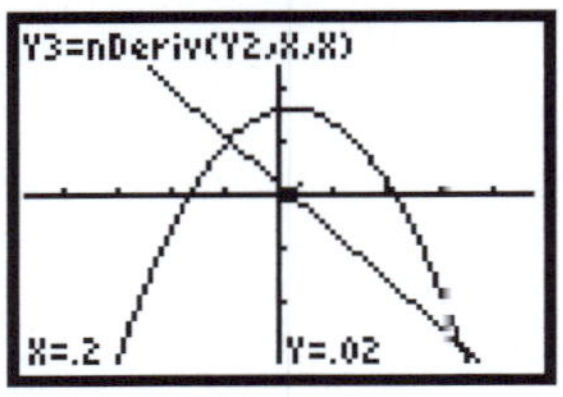

Using the Table with successively smaller values of **ΔTbl** suggests a value of X of about 0.222.

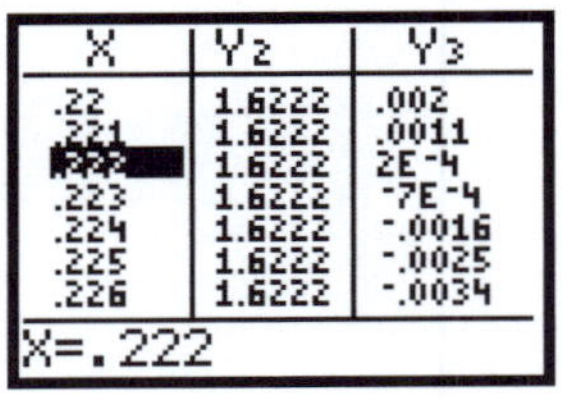

(b)

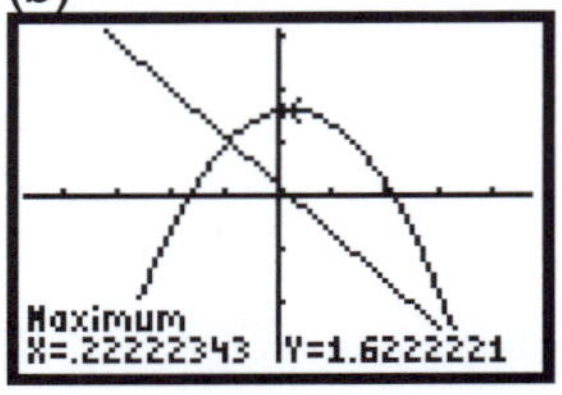

(c)

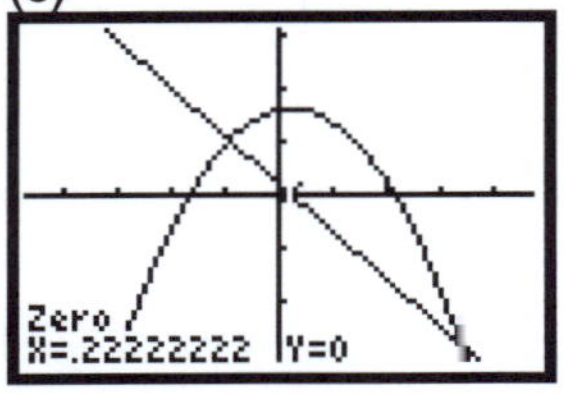

(d)

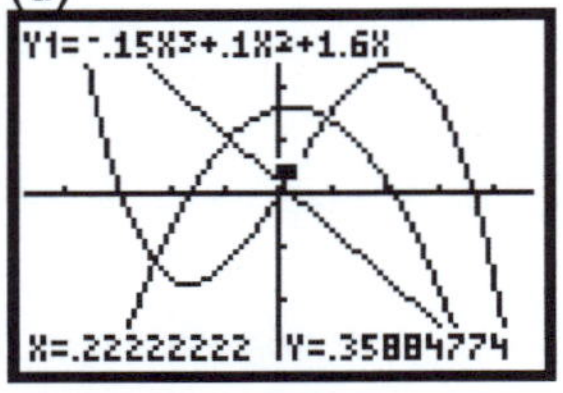

(3) (a) For this particular function, at the point of inflexion, the gradient is positive, but it could equally well be negative – however, it would still be steepest there. Recall the distinction that was drawn in Unit 1 between the terms *steepness* and *gradient.*

(4) (a) Since $\frac{dy}{dx} = {}^{-}0.45x^2 + .2x + 1.6$, $\frac{d^2y}{dx^2} = {}^{-}0.9x + .2$.

(b) Putting $\frac{d^2y}{dx^2} = 0$, gives ${}^{-}0.9x + .2 = 0$ so $x = 2/9$, confirming the values calculated in Activity 2.

(5) The methods of Activity 2 each lead to a value of about -1.3333 for the point of inflexion.

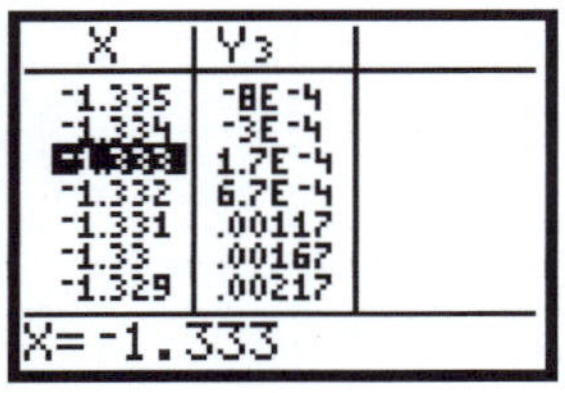

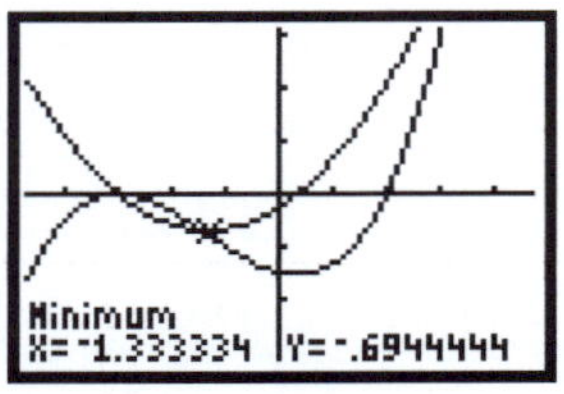

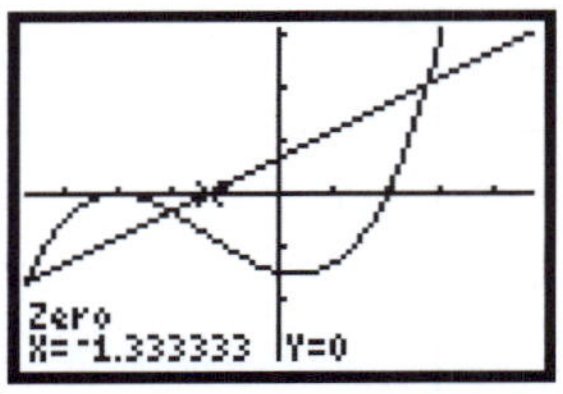

Since $y = x^3/12 + x^2/3 - x/4 - 1.5$, $\frac{dy}{dx} = 3x^2/12 + 2x/3 - 1/4$ and $\frac{d^2y}{dx^2} = 6x/12 + 2/3 = x/2 + 2/3$.

Putting $x/2 + 2/3 = 0$ gives $x = {}^{-}4/3$, and this confirms the values calculated above.

(6) There are two points of inflexion. The derived function is a cubic and the second derived function a quadratic. The quadratic cuts the x-axis at two points whose x-coordinates correspond to those of the points of inflexion of the original quartic function.

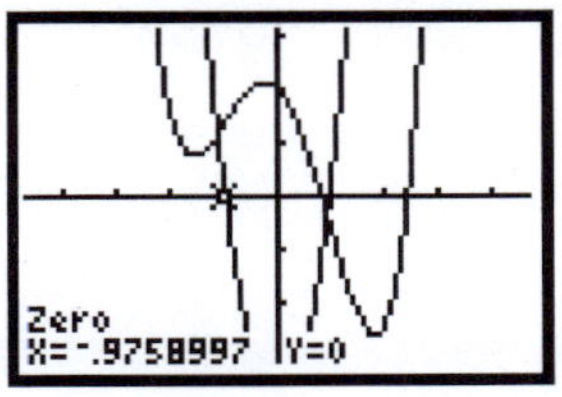

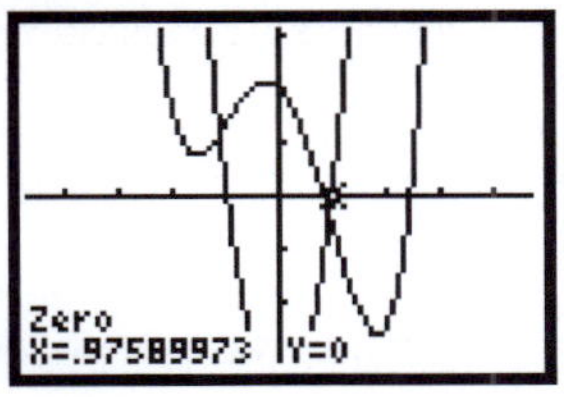

Using **zero** from the Calc menu on Y3 suggests that the x-coordinates of the points of inflexion are about ±0.9759.

Since $y = .35x^4 - 2x^2 - x + 2$, $\frac{dy}{dx} = 1.4x^3 - 4x - 1$ and $\frac{d^2y}{dx^2} = 4.2x^2 - 4$.

Putting $\frac{d^2y}{dx^2}=0$, gives $4.2x^2 - 4 = 0$ so $x = \sqrt{(4/4.2)} = \pm 0.9759$ (to 4 d.p), confirming the values above.

(7)

PROGRAM: INFLEX	[PRGM] [◄] **1 INFLEX** [ENTER]
:"nDeriv(Y1,X,X)" → Y2	[ALPHA] **"** [MATH] **8** [VARS] [►] **1 1** [,] [X,T,Θ,n] [,] [X,T,Θ,n] [)] [ALPHA] **"** [STO►] [VARS] [►] **1 2** [ENTER]
:"nDeriv(Y2,X,X)" →Y3	[ALPHA] **"** [MATH] **8** [VARS] [►] **1 2** [,] [X,T,Θ,n] [,] [X,T,Θ,n] [)] [ALPHA] **"** [STO►] [VARS] [►] **1 3** [ENTER]
:FnOff 2,3	[VARS] [►] **4 2 2** [,] **3** [ENTER]
:Text(6,0,"GUESS?")	[2nd] [DRAW] **0 6** [,] **0** [,] [2nd] [A-LOCK] **"GUESS?"** [ALPHA] [)] [ENTER]
:Trace	[TRACE] [ENTER]
:X→G	[X,T,Θ,n] [STO►] [ALPHA] **G** [ENTER]
:Text(0,0,"POINT OF INFLEXION")	[2nd] [DRAW] **0 0** [,] **0** [,] [2nd] [A-LOCK] **"POINT␣OF␣INFLEXION"** [ALPHA] [)] [ENTER]
:Text(6,0,"AT X=", solve(Y3,X,G))	[2nd] [DRAW] **0 6** [,] **0** [,] [2nd] [A-LOCK] **"AT␣X** [2nd] [TEST] **1** [ALPHA] **"** [,] [MATH] **0** [VARS] [►] **1 3** [,] [X,T,Θ,n] [,] [ALPHA] **G** [)] [)]

Press [2nd] [QUIT] to return to the home screen.

Executing the program twice (once for each point of inflexion) produces these results, confirming the previous x-coordinates.

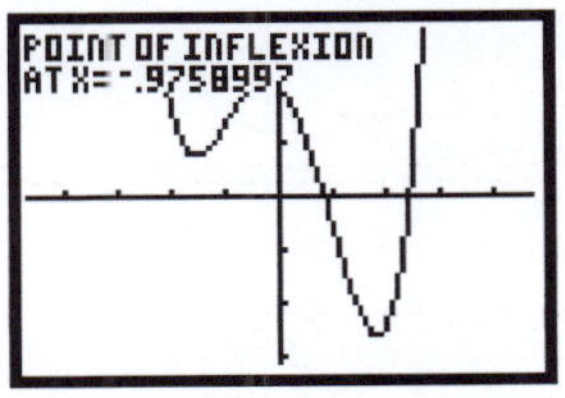

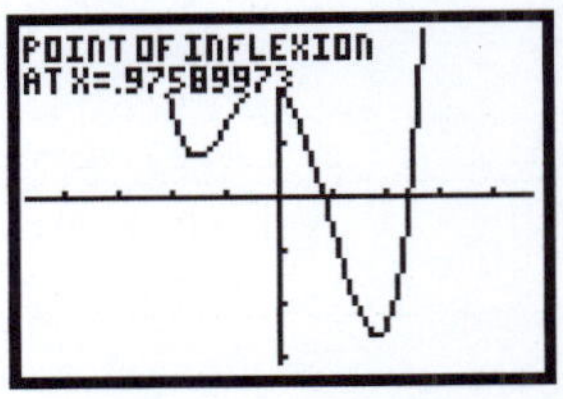

(8) All answers are given correct to 4 d.p.
(a) X = 0.3333 (b) X = ±1.1547
(c) X = -1.0000 and X = 1.5000
(d) X = 0 and X = ±2.4495.

Brain Stretcher

In the case of **Y1=X+2** the second derived function is Y3=0, so in a sense there are points of inflexion at every point on the graph – not very sensible!

In the case of **Y1=X²–2** the second derived function is Y3=2, and this does not cross the x axis, so there should be no points of inflexion. Some quirk in the calculator produces a large negative number, which is clearly wrong.

A way to fix the bug in the program is to check whether Y3 is a horizontal line. This can be done for example by checking whether Y3(0) is the same as Y3(1). If it is, a suitable message can be displayed.

The full program is shown here.

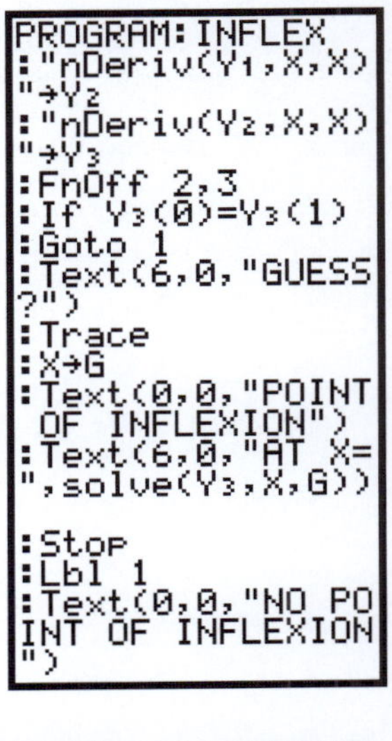

Unit 10 Rates of change

(1) (a)

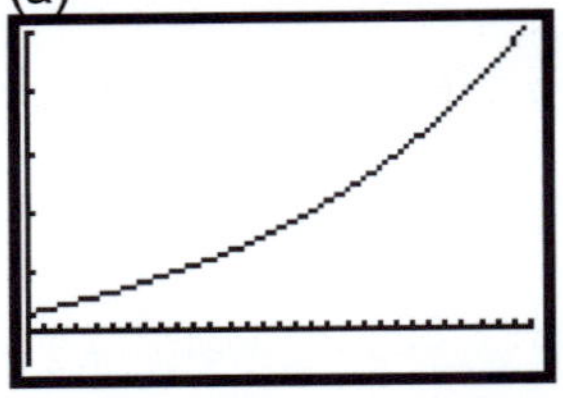

X	Y1
0	30
5	66.25
10	110
15	168.75
20	250
25	361.25
30	510

X=0

The costs are in the Y1 column.

(b) The marginal costs can be found in two ways; either by repeated use of **dy/dx** (Option 6 in the CALC menu), or by entering **Y2=nDeriv(Y1,X,X)** and using the Table again.

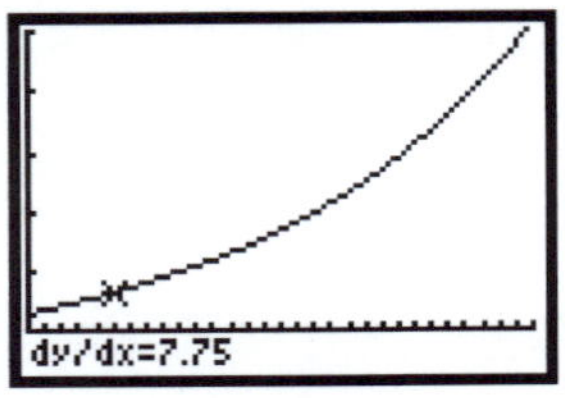

X	Y1	Y2
0	30	7
5	66.25	7.75
10	110	10
15	168.75	13.75
20	250	19
25	361.25	25.75
30	510	34

X=0

(2) (a)

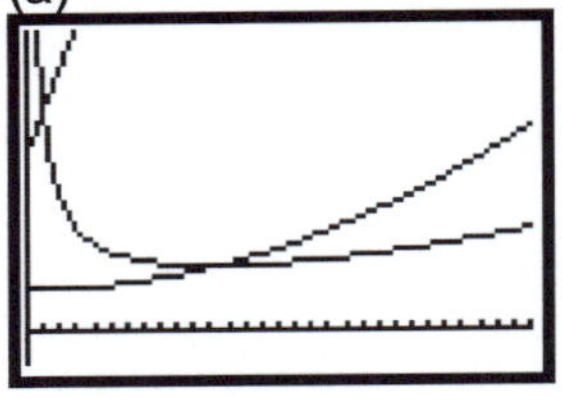

While marginal costs gradually increase, mean costs initially decline before starting to rise again.

(b)

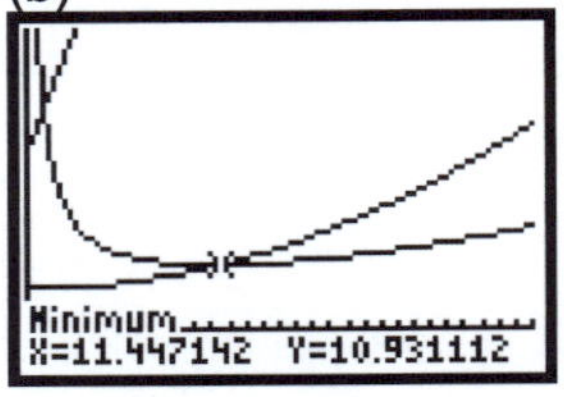

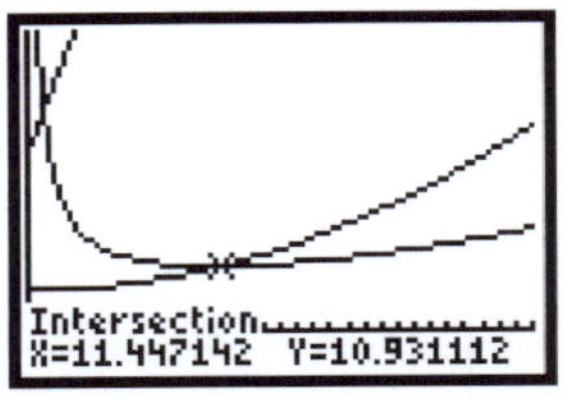

(3) (a)

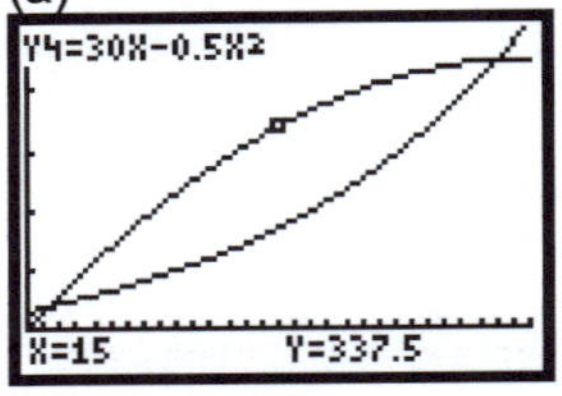

(b)
X=15 is a good estimate for the level of production which creates maximum profit because the two graphs seem to be furthest apart about there. Using Option 4 in the CALC menu gives a maximum profit of just over 169 at about X=15.65.

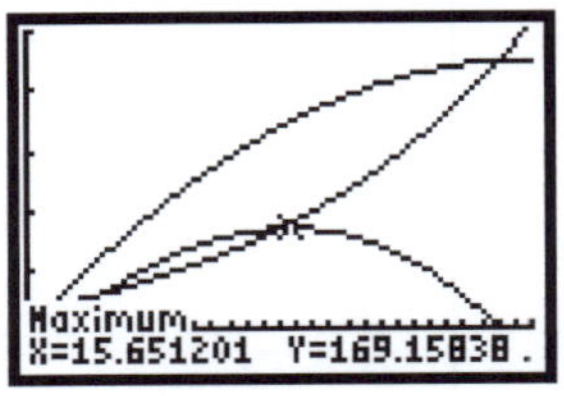

(c)
Since actual profit is in Y5, entering **Y6= nDeriv(Y5,X,X)** allows marginal profit to be calculated. Using the Table produces the required values as shown here.

X	Y5	Y6
0	-30	23
5	71.25	17.25
10	140	10
15	168.75	1.25
20	150	-9
25	76.25	-20.75
30	-60	-34

X=0

(5) (a)

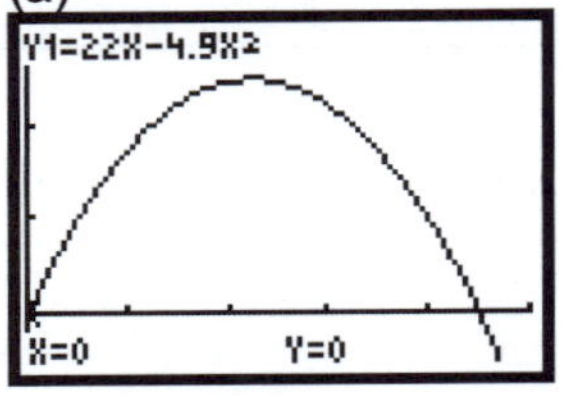

A bystander would see the arrow going straight up and down, not moving in a parabolic shape as on this graph. It is important to remember that what is plotted on the horizontal axis is time, not horizontal distance. However, it is possible to see a more realistic view of the flight of the arrow, (including a moving image on the calculator screen) using Parametric Graphing. See the Brain stretcher at the end of this activity in the main text.

The questions can be answered using options 1, 2, 4 and 6 of the CALC menu.

(b) Use option 1 (or Trace).

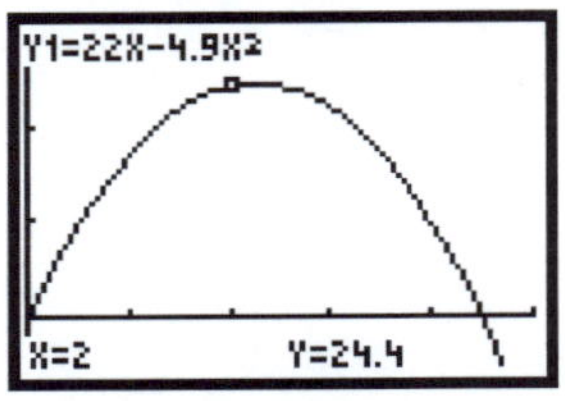

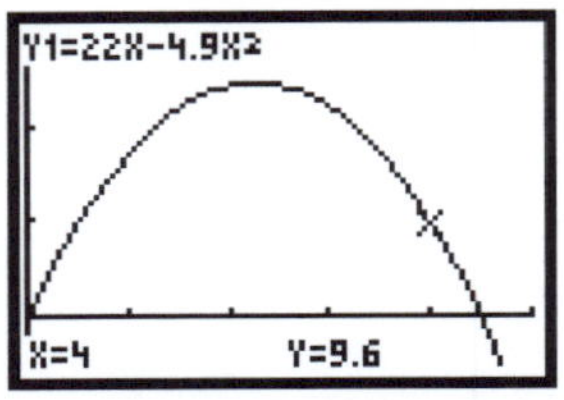

(c) Use option 6:**dy/dx**.

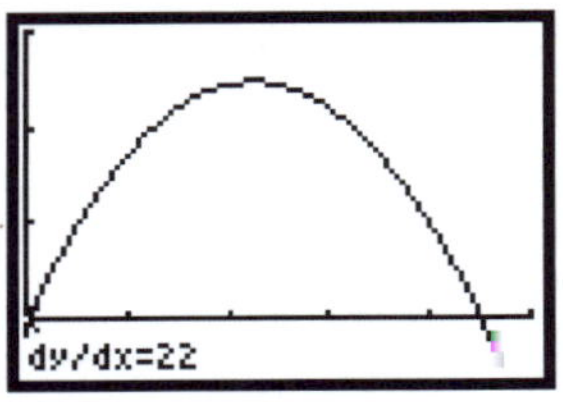

(d) Use option 2: **value**.

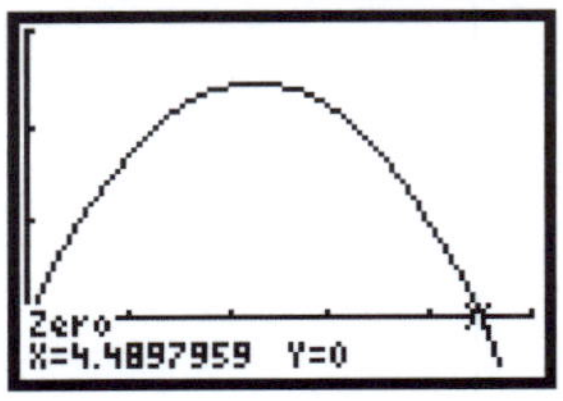

(e) Use option 6 and enter **Ans** for the X coordinate from part (d) Compare with part (c): this is the same speed, but the opposite direction.

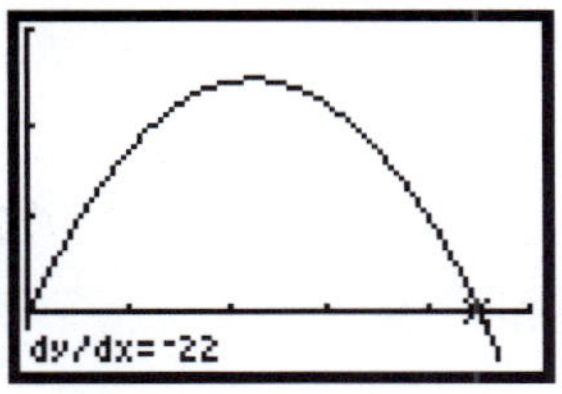

(f) Use option 4: **maximum**. (Your answers may vary very slightly from the ones shown here – it depends on the choice of window settings.)

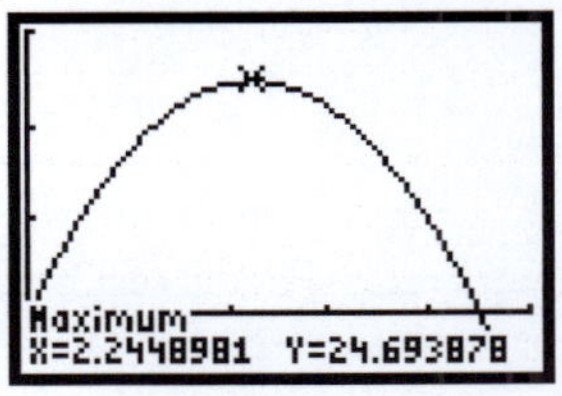

(g) Use option 6 and enter **Ans** for the X coordinate from part (f). The calculated value is very close to zero, the true velocity at the highest point.

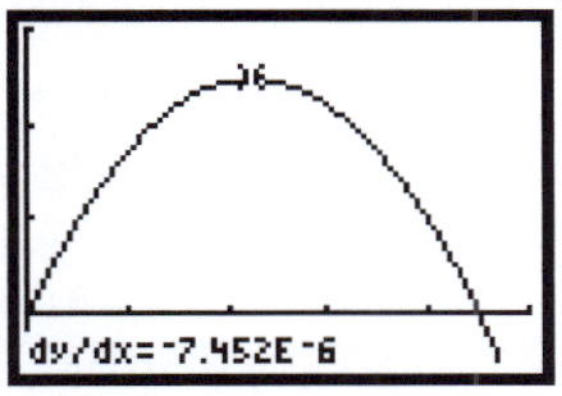

(h) The graph shows that the velocity is positive to begin with (the arrow is going up) and then becomes negative (as the arrow comes down). The straight line indicates a constant change in the velocity, that is, constant acceleration.

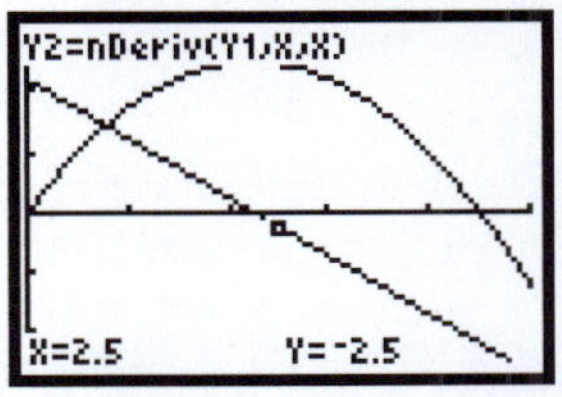

(i)

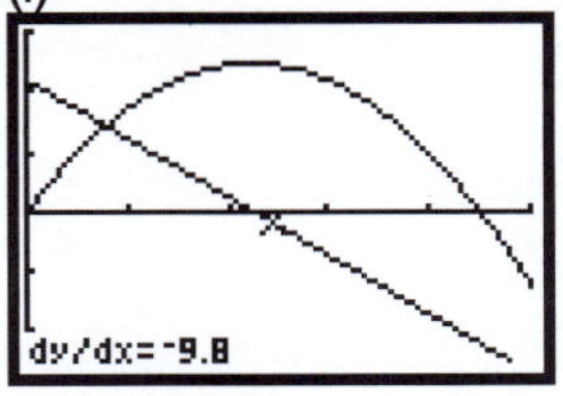

The value of *dy*/*dx* is ⁻9.8 at every point on the velocity time graph. This is (approximately) the acceleration due to gravity: ⁻9.8 m/sec^2.

(6) (a)

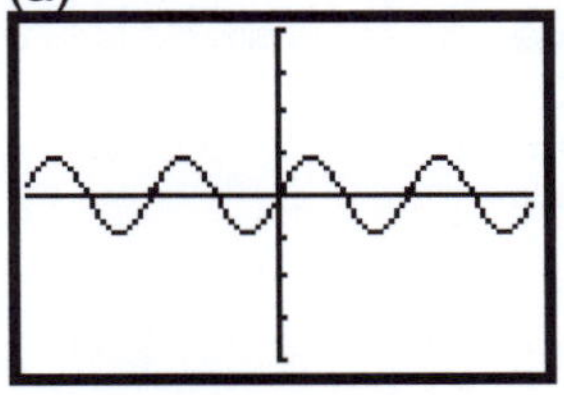

(b) *A* determines the height of the wave-form – it is known as the amplitude of the graph. *W* determines the frequency with which the wave oscillates. *F* moves the wave along the horizontal axis – it is known as the phase difference. For example, a value of *F*=π/2 moves the original graph a horizontal distance of π/2.

(c) Y2 represents the velocity function and Y3 the acceleration function.

(d)

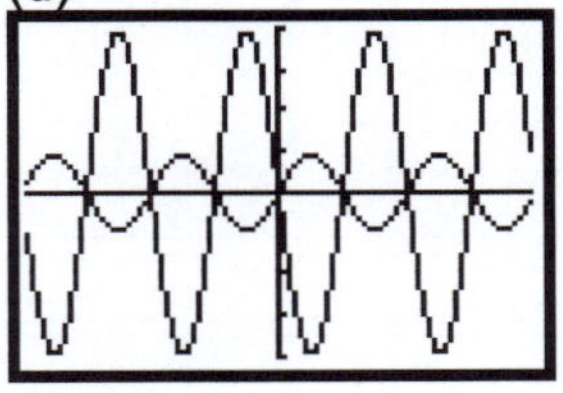

(e) In this case ω = 2, so the expected relationship is: acceleration = ⁻4 × displacement, or Y3= ⁻4×Y1. This relationship appears to be true if the graphs are compared, perhaps using Trace. A more convincing demonstration is to use the Table as shown here with values of **ΔTbl** of π/8.

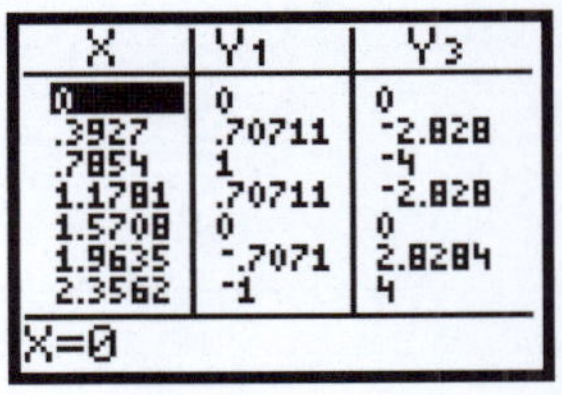

X	Y1	Y3
0	0	0
.3927	.70711	⁻2.828
.7854	1	⁻4
1.1781	.70711	⁻2.828
1.5708	0	0
1.9635	⁻.7071	2.8284
2.3562	⁻1	4

X=0

Unit 11 Antiderivatives

(1) The function that was actually used was **Y1=1/(1+X²)–1**, as shown here. However it might equally well have been **Y1=1/(1+X²)+5** or **Y1=1/(1+X²)–3** etc. The value of the constant term does not affect the value of the gradient function.

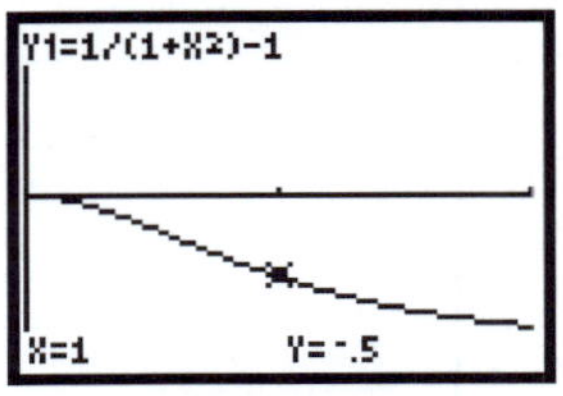

(2) In each case, one possible function is shown below, but there are many other possibilities.

(a) Since Y2 is the horizontal line Y=1, it follows that the gradient of Y1 is constant and equal to 1, So Y1 could be X or X+1, X–2, etc.

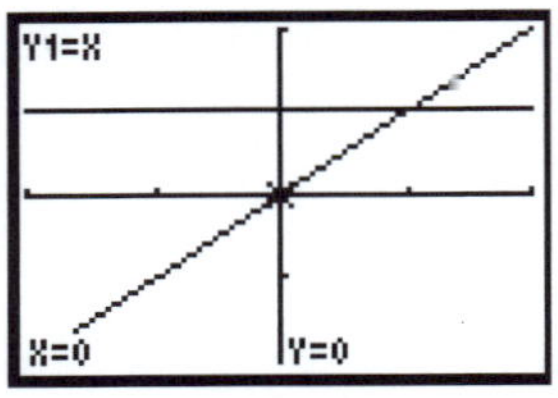

(b) Since Y2 is the horizontal line Y=$^-0.5$, it follows that the gradient of Y1 is constant and equal to $^-0.5$. So Y1 could be $^-0.5$X or $^-0.5$X+1, $^-0.5$X–2, etc.

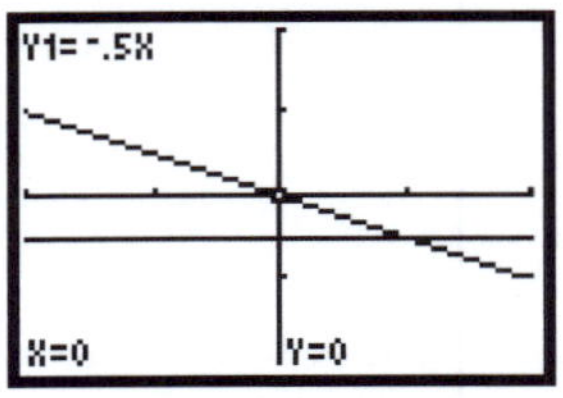

(c) Since Y2 is the straight line Y=2X, it follows that the gradient of Y1 increases steadily. This gradient starts as a negative value, is zero at X=0 and then becomes positive. So **Y1=X²** is one possible solution.

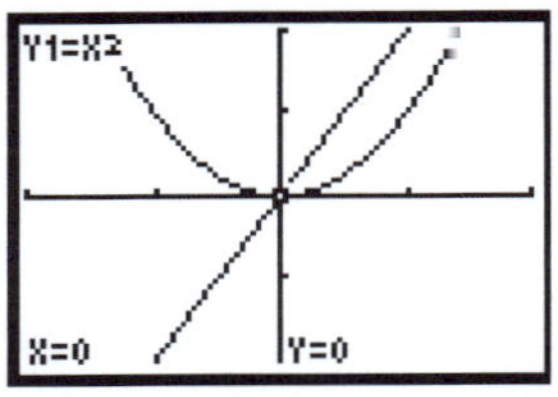

(d) This is much more difficult. The gradient of Y1 is negative until about X=$^-0.7$ where it is zero. It then becomes positive for values of X up to about 0.7, before becoming negative again. So **Y1=X–X³** is one possible solution.

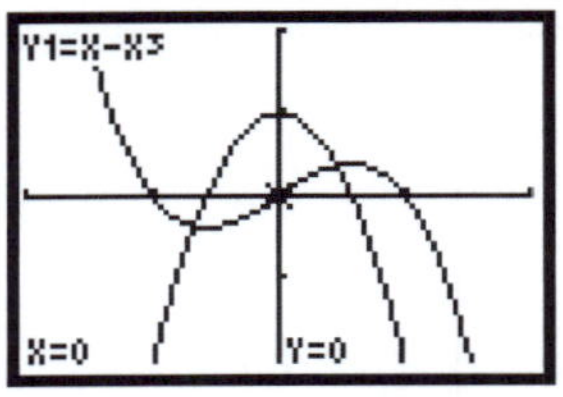

(5) (a)

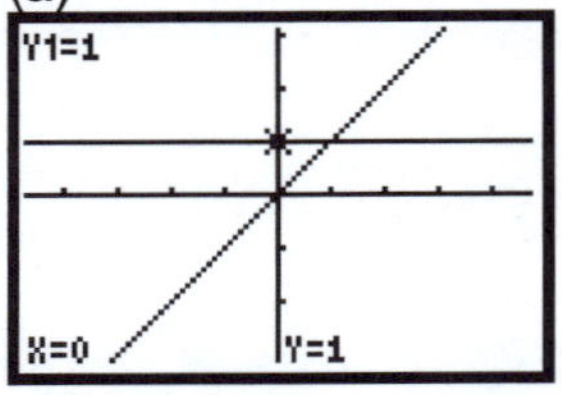

(b)

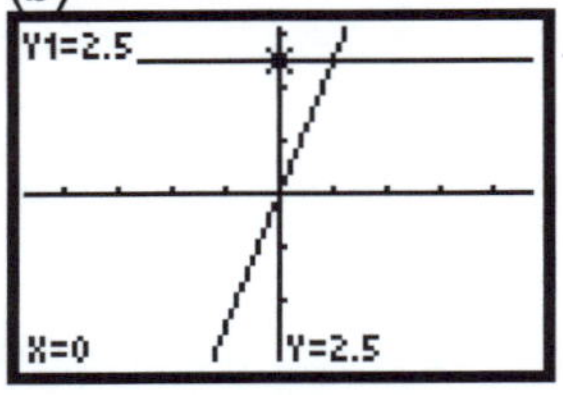

(c)

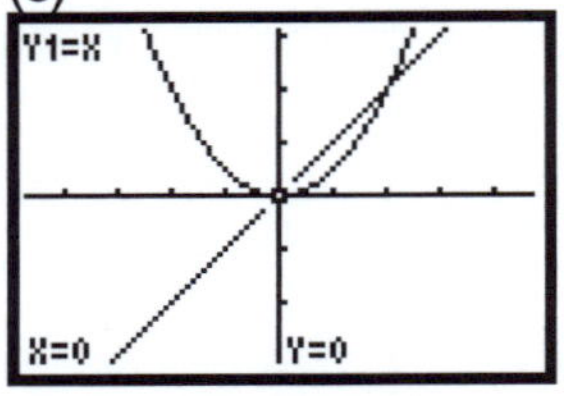

(d)

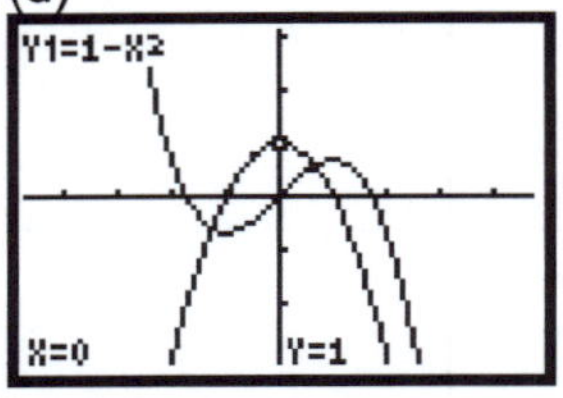

(6) Whatever function you enter in Y1, its graph will be the same as that of Y3.

(10)

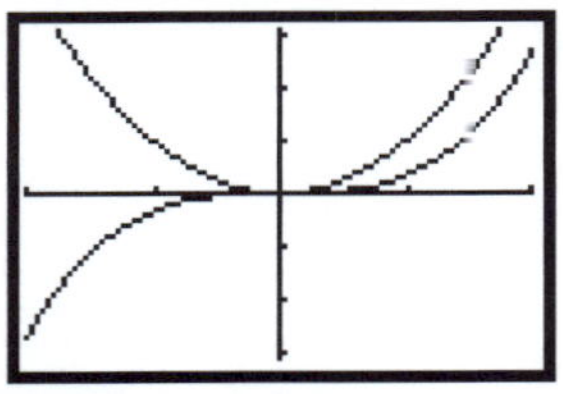

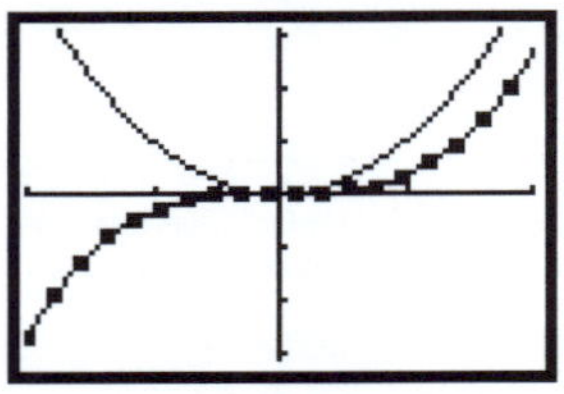

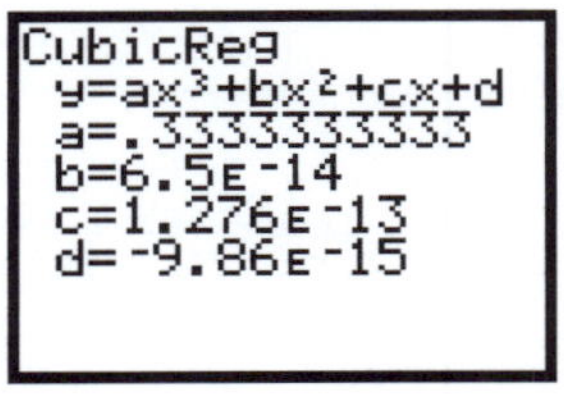

Treating the extremely small values for **b**, **c** and **d** as zero, it seems that the equation of the function in Y2 is indeed $y=x^3/3$.

So this confirms that $\int x^2\,dx = \frac{x^3}{3}$ + some constant of integration.

Brain stretcher

The general rule holds for n = 1 – see Activity 5(c), and also for n = 0 – see Activity 5(a). (Note: x^0 = 1). However, it does not hold for n = $^-1$:

if you try to use the rule you get $\int x^{-1}\,dx = \frac{x^0}{0} - c$ and division by zero is not possible. Similarly, if you try to use **fnInt(Y1,X,0,X)** the calculator produces an error message.

(11) (a) Using the general rule: $\int(4x^2 + 3x + 1)\ dx = \frac{4x^3}{3} + \frac{3x^2}{2} + x + c$.
Y2=fnInt(Y1,X,0,X) and **Y3=4x³/3+3X²/2+X** produce the same graph.

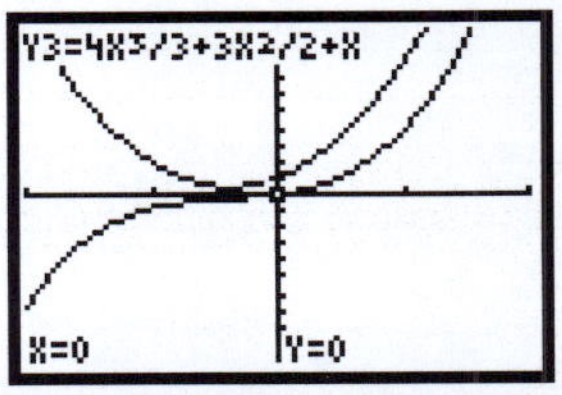

Alternatively using program **SCATTER** and cubic regression produces these results.

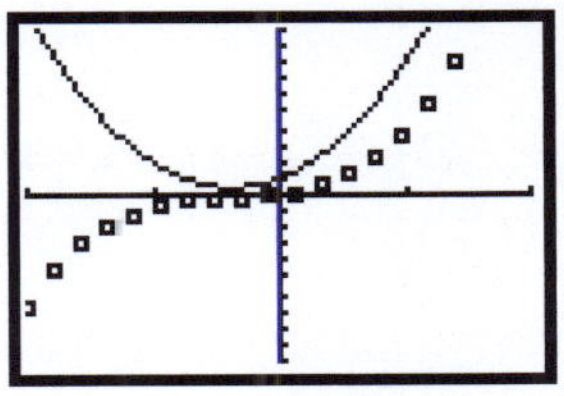

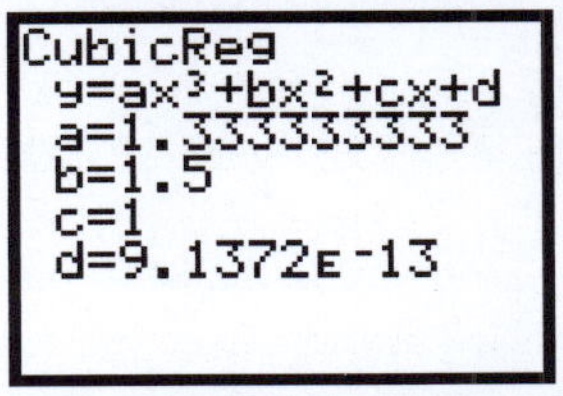

(b) Using the general rule: $\int(4x^3 - 2x)\ dx = \frac{4x^4}{4} - \frac{2x^2}{2} + c = x^4 - x^2 + c$.
Y2=fnInt(Y1,X,0,X) and **Y3=X^4–X²** produce the same graph.

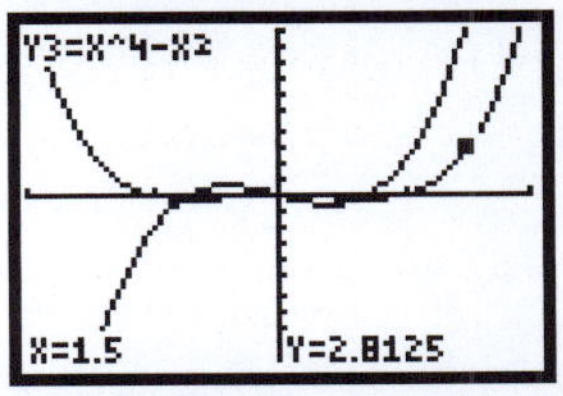

Alternatively using program **SCATTER** and quartic regression produces these results.

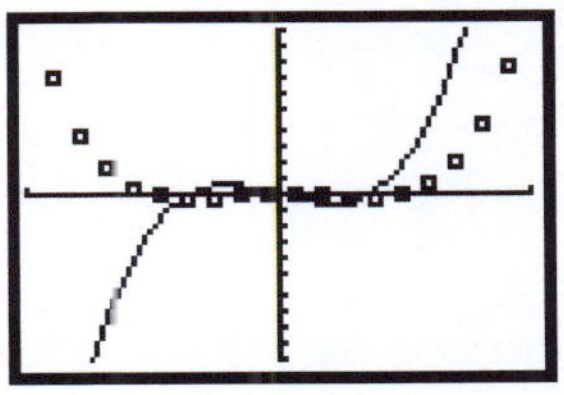

(c) Using the general rule $\int x^{\frac{1}{2}}dx = x^{\frac{3}{2}} \quad \frac{3}{2} + c = \frac{2x^{\frac{3}{2}}}{3} + c$.

Y2=fnInt(Y1,X,0,X) and **Y3=2X^(3/2)/3** produce the same graph (using the window settings suggested).

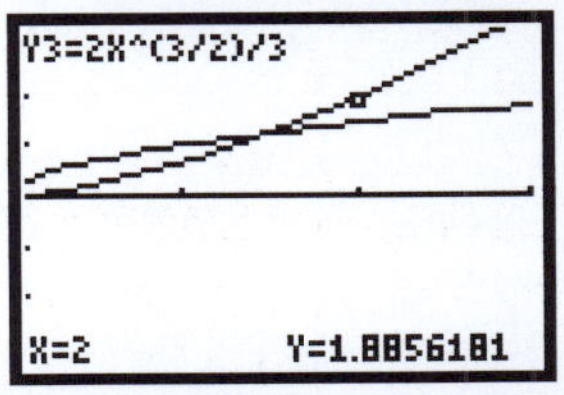

Alternatively using program **SCATTER** and power regression produces these results.

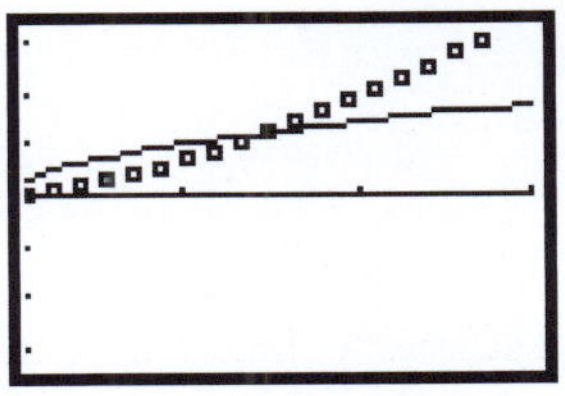

Brain stretcher

Try entering a simple function which has a constant term such as $y = x^2+2$. When you differentiate to form Y2, the constant term is lost and there is no way that Y3 can recreate it. However, the same problem does not arise when you integrate first and then differentiate, as in Activity 6.

So finally …

◊ The indefinite integral of a constant is a linear function, of a linear function is a quadratic function, of a quadratic function is a cubic function, and of a cubic function is a quartic function, and so on for all polynomial functions.

◊ When you calculate a derivative, all constant terms disappear, so when you calculate an antiderivative there is no way of knowing which, if any, constants are needed. So when you carry out indefinite integration it is usual to add a constant of integration, c, as a reminder that some unknown constant is probably a part of the integral function.

Unit 12 Definite integrals

(1) You will no doubt have found that it is useful to use **ClrDraw** (Option 1 in the Draw menu) before using **∫f(x)dx** for a second time. The value that is calculated is discussed in the following activities.

(2) 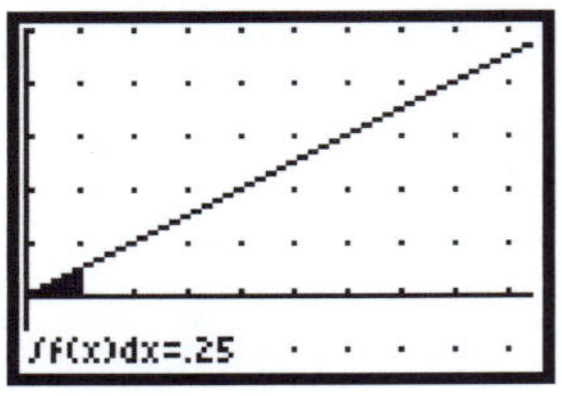

(a)

The calculated value of **∫f(x)dx** is equal to the shaded area on the graphing screen: that is the area beneath the graph of the function $y = x/2$, and between X=0 and X=1.

(b)

Lower limit	0	0	0	0	0	0	0	0	0
Upper limit	1	2	3	4	5	6	7	8	9
∫f(x)dx	.25	1	2.25	4	6.25	9	12.25	16	20.25

(c) The values of **∫f(x)dx** are equal to the upper limit squared and divided by 4.

If the upper limit were 10, **∫f(x)dx** would be $10^2 \div 4 = 25$.
If the upper limit were 2.5, **∫f(x)dx** would be $2.5^2 \div 4 = 1.5625$.
If the upper limit were *N*, **∫f(x)dx** would be $N^2 / 4$.

(d) $\int x/2 \, dx = x^2/4 + c$. There are clear similarities with $N^2 / 4$, with *N* replaced by *x* and a constant of integration, *c*, added.

(e) For the function $y = x/2$, the calculator command **∫f(x)dx**, if applied between a lower limit of 0 and an upper limit of *N*, produces a value equal to the area under the graph of the function $y = x/2$, between X=0 and X=*N*. This value is the same as produced by finding the antiderivative of $y = x/2$, setting the constant $c=0$ and substituting $x=N$.

(3) (a) 2.75 (b) 8 (c) 1.0625

(5) (a) There are nine squares between the vertical lines. It is possible to do the area-of-a-trapezium calculation on the home screen as shown here, but it is much easier to do it in your head – $(2+4)/2\times3 = 9$. Using **∫f(x)dx** confirms this result as shown.

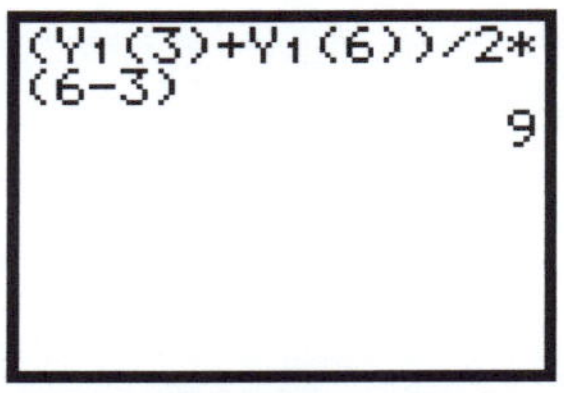

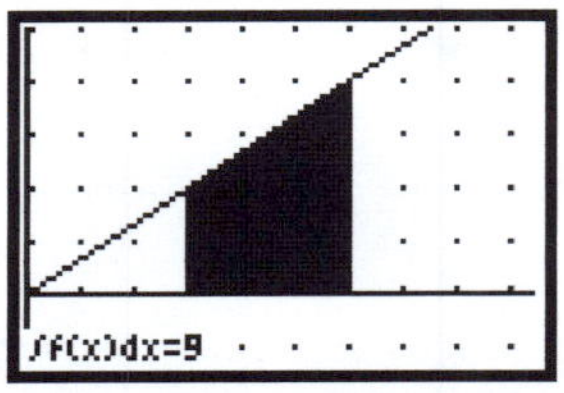

(b) Counting squares is not really an option this time and the home-screen calculation method comes into its own with these more awkward X values.

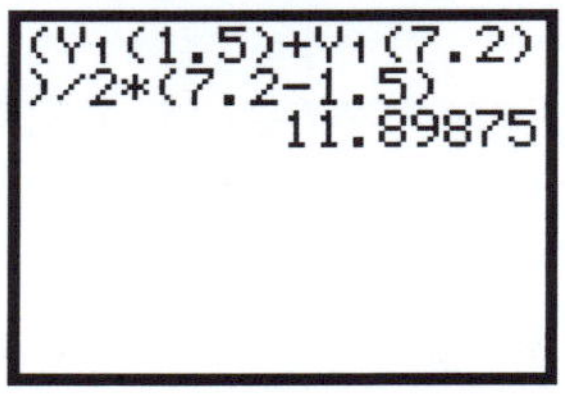

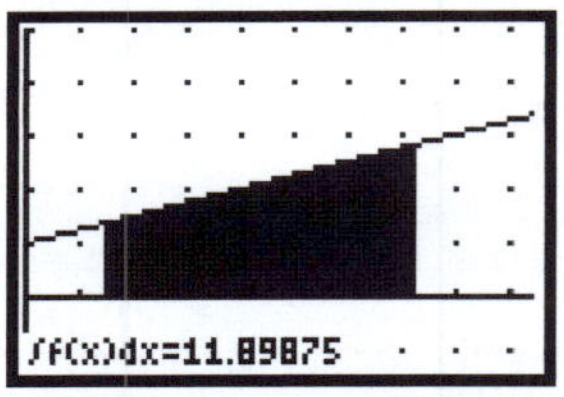

(c) A rough estimate (counting the number of squares) might give an area of 9.5 or 10 square units.

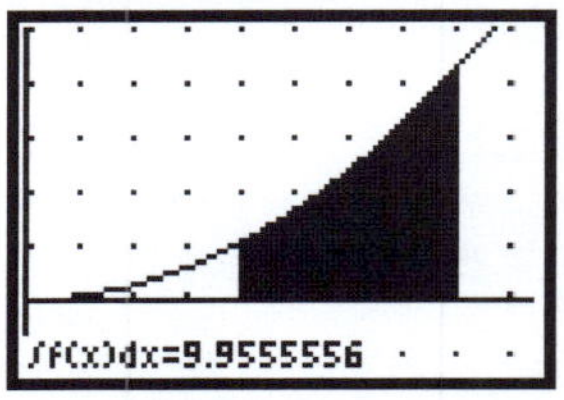

(7) (a) $\int_1^2 \frac{x^2}{15} dx = \left[\frac{x^3}{45}\right]_1^2 = \frac{2^3}{45} - \frac{1^3}{45} = 0.156$ to 3 d.p.

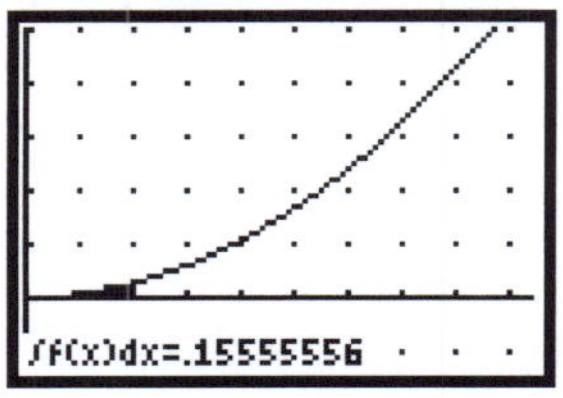

(b) $\int_2^4 (\frac{x^2}{15}+\frac{x}{4})dx = \left[\frac{x^3}{45}+\frac{x^2}{8}\right]_2^4 = \frac{4^3}{45}+\frac{4^2}{8}-\frac{2^3}{45}-\frac{2^2}{8} = 2.744$ to 3 d.p.

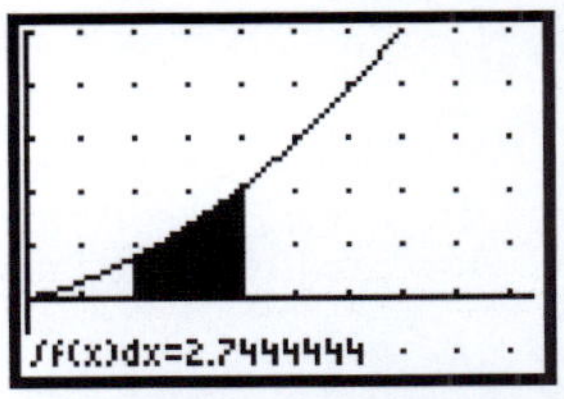

(c) $\int_0^8 (2x-\frac{x^2}{4})dx = \left[x^2-\frac{x^3}{12}\right]_0^8 = 8^2-\frac{8^3}{12} = 21.333$ to 3 d.p.

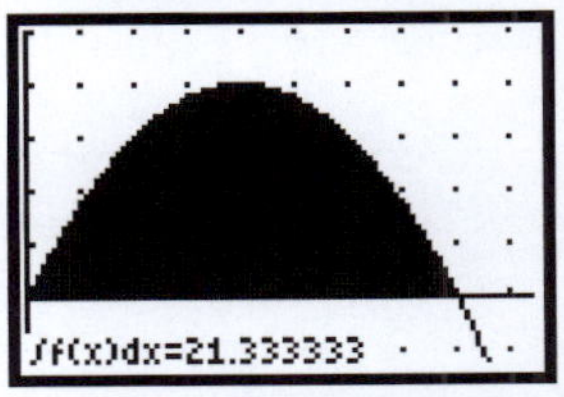

(d) $\int_1^5 x^{\frac{1}{2}} dx = \left[x^{\frac{3}{2}} \quad \frac{3}{2}\right]_1^5 = \frac{2}{3}\left[x^{\frac{3}{2}}\right]_1^5 = \frac{2}{3}\left[5^{\frac{3}{2}}-1^{\frac{3}{2}}\right] = 6.787$ to 3 d.p.

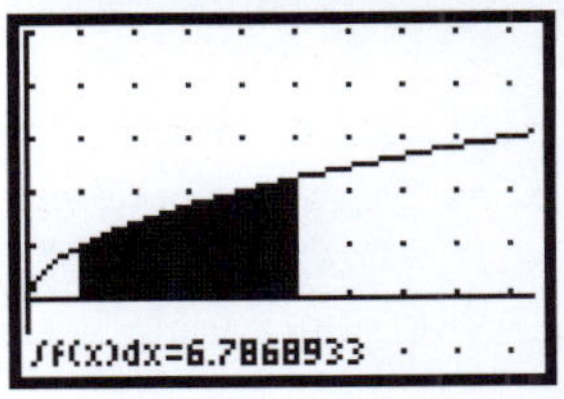

(8)

```
fnInt(X²/15,X,1,
2)
          .1555555556
fnInt(X²/15+X/4,
X,2,4)
          2.744444444
```

```
fnInt(2X-X²/4,X,
0,8)
          21.33333333
fnInt(X^(1/2),X,
1,5)
          6.786893258
```

Unit 13

(2) (a) The key presses required for program **TRAPRULE** are below.

PROGRAM: TRAPRULE	[PRGM] [◄] **1 TRAPRULE** [ENTER]
:ClrDraw	[2nd] [DRAW] **1** [ENTER]
:0→A	**0** [STO►] [ALPHA] **A** [ENTER]
:Prompt W	[PRGM] [►] **2** [ALPHA] **W** [ENTER]
:For(C,W,1,W)	[PRGM] **4** [ALPHA] **C** [,] [ALPHA] **W** [,] **1** [,] [ALPHA] **W** [)] [ENTER]
:A+(Y1(C–W)+Y1(C))/2*W→A	[ALPHA] **A** [+] [(] [VARS] [►] **1 1** [(] [ALPHA] **C** [–] [ALPHA] **W** [)] [+] [VARS] [►] **1 1** [(] [ALPHA] **C** [)] [)] [÷] **2** [×] [ALPHA] **W** [STO►] [ALPHA] **A** [ENTER]
:Line(C,0,C,Y1(C))	[2nd] [DRAW] **2** [ALPHA] **C** [,] **0** [,] [ALPHA] **C** [,] [VARS] [►] **1 1** [(] [ALPHA] **C** [)] [)] [ENTER]
:Line(C,Y1(C),C–W,Y1(C–W)	[2nd] [DRAW] **2** [ALPHA] **C** [,] [VARS] [►] **1 1** [(] [ALPHA] **C** [)] [,] [ALPHA] **C** [–] [ALPHA] **W** [,] [VARS] [►] **1 1** [(] [ALPHA] **C** [–] [ALPHA] **W** [)] [ENTER]
:End	[PRGM] **7** [ENTER]
:Pause	[PRGM] **8** [ENTER]
:Disp A	[PRGM] [►] **3** [ALPHA] **A**

Finally, press [2nd] [QUIT] to return to the home screen.

(b) Repeated use of the program gives results like those shown here. It seems that the values of A are tending towards 0.3333...

```
                 Done
W=?.05
              .33375
                 Done
W=?.02
               .3334
                 Done
```

(c) Repeated use of the program gives results like those shown here. It seems that the values of A are tending towards 0.6666...

```
                 Done
W=?.05
                .665
                 Done
W=?.02
               .6664
                 Done
```

(3) The command **∫f(x)dx** produces a more accurate answer, much more quickly.

(4)(a) The area shaded is certainly not zero, despite the result at the bottom of the screen!

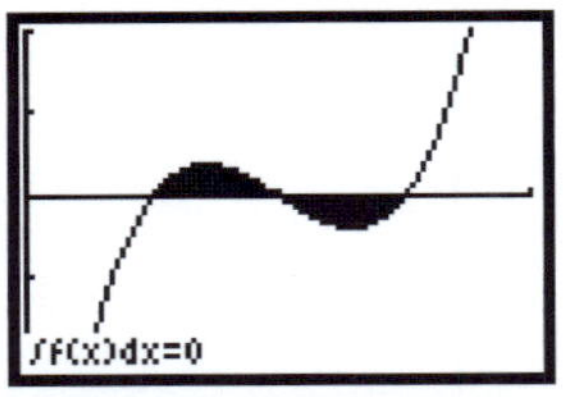

(b) Either using **fnInt** on the home screen, or **∫f(x)dx** on the graphing screen gives the same results: 0.25 and ⁻0.25. In part (a) these two values have been added together to produce zero.

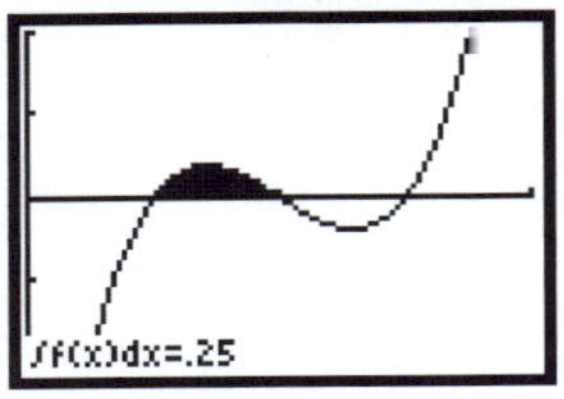

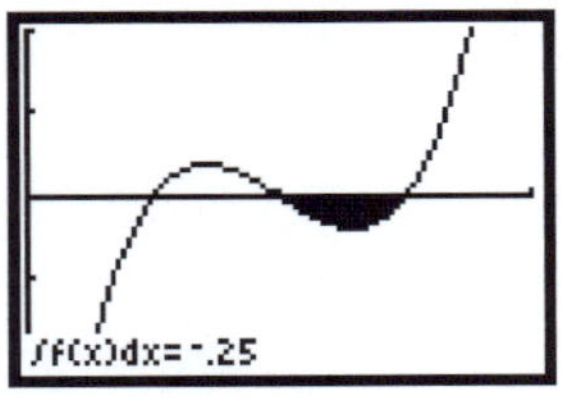


```
fnInt((X-1)(X-2)
(X-3),X,1,2)
                .25
fnInt((X-1)(X-2)
(X-3),X,2,3)
               -.25
```

(c) Expanding $(x–1)(x–2)(x–3)$ gives $x^3–6x^2+11x–6$.

Note: you can check such an expansion by entering the two forms as Y1 and Y2 and pressing [2nd] [TABLE] to ensure that the same values are tabulated.

$$\int_1^3 (x^3 - 6x^2 + 11x - 6)dx = \left[\frac{x^4}{4} - 2x^3 + \frac{11x^2}{2} - 6x\right]_1^3 = (^-2.25) - (^-2.25) = 0.$$

(5) (a) The definite integral is zero but the area is clearly not.

There are 4 sq. units below the axis and 4 sq. units above. Total area = 8 sq. units.

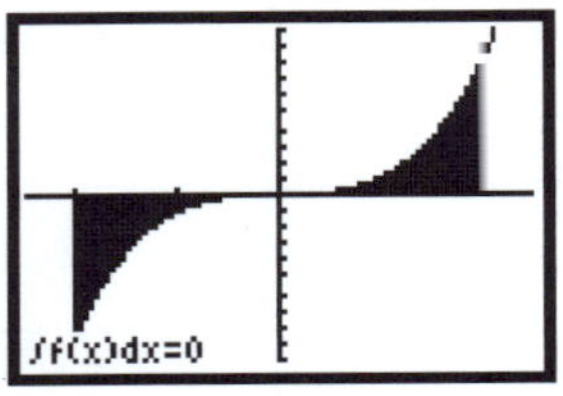

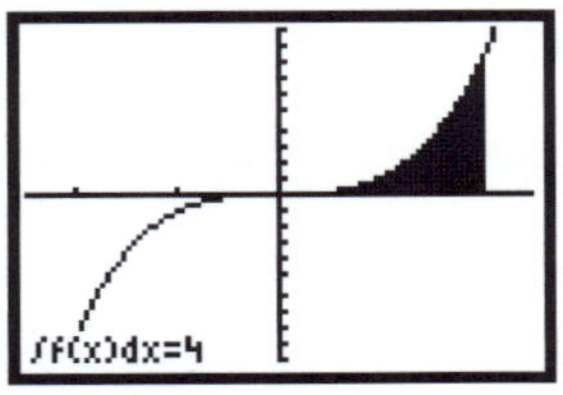

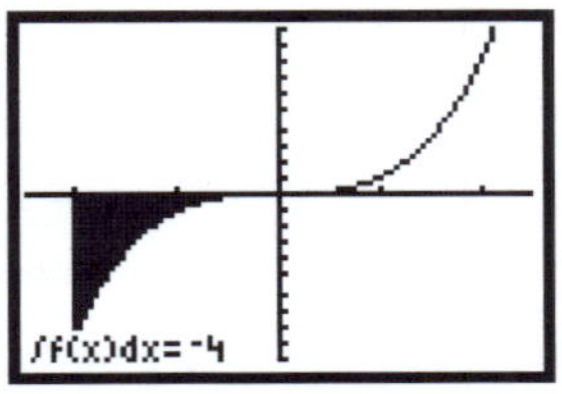

(b) The definite integral is ⁻0.416 but since the graph lies below the curve the area is actually 0.416 sq.units.

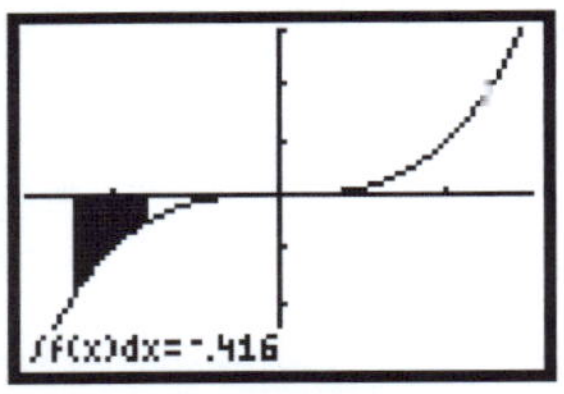

(c) The definite integral is 1.015625 but since some of the graph lies above and some below the axis, the area must be calculated in two parts.

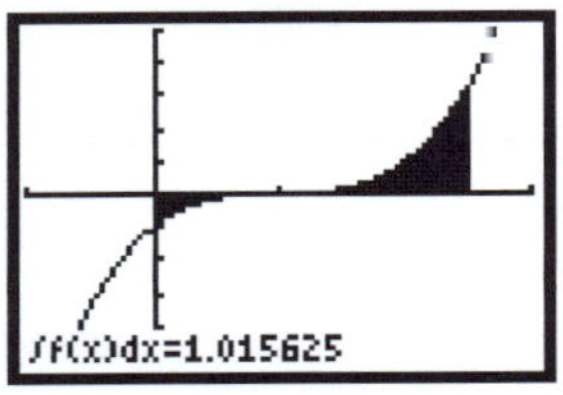

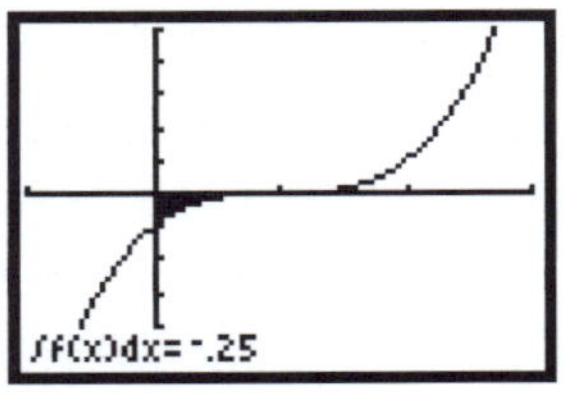

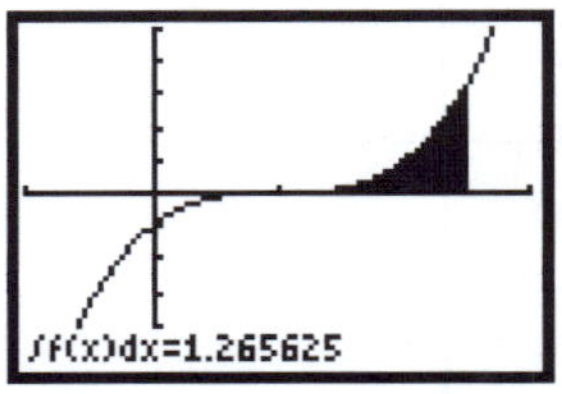

The curve crosses the axis at x=1. There is 0.25 sq. units below the axis and 1.265625 sq.units above. Total area = 1.52 sq.units (to 2 d.p.)

(d) Drawing the graph shows one area above and one below the axis and the symmetry of the sine curve suggests that the two areas are equal. The curve obviously cuts the x-axis at X=π.

The total area must be 4 sq. units.

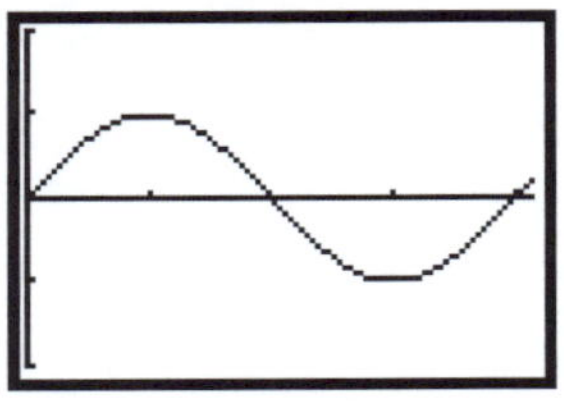

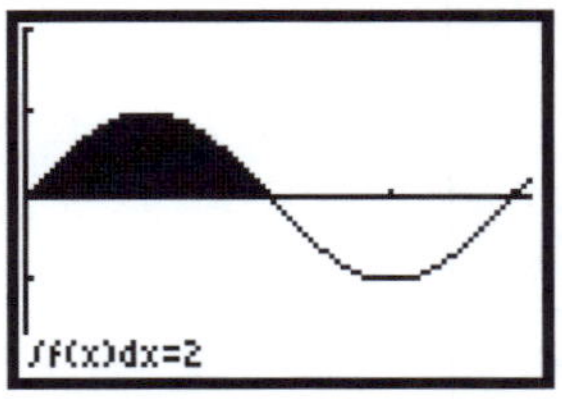

(6) You probably did not slog through the long method to calculate the complex area, but you may be interested to see some of the steps along the way when we tried this method, to find that the area was 5.060 (to 3 d.p.). The second screen shows confirmation that the graph is symmetrical about the value x=π.

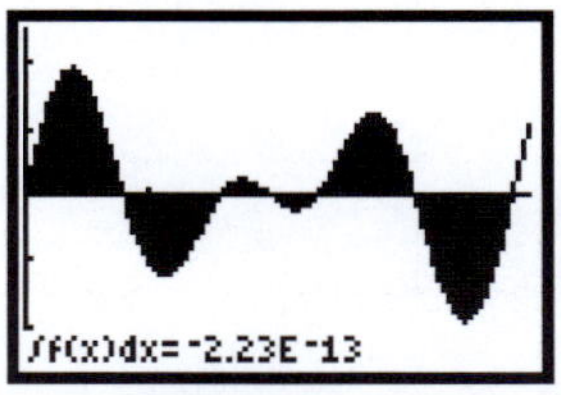

X	Y1
2.1991	-.642
2.5133	0
2.8274	.22123
3.1416	0
3.4558	-.2212
3.7699	0
4.0841	.64204

Y1=-.64203952192

```
Ans÷A
           1.256637061
Ans÷B
           2.513274123
```

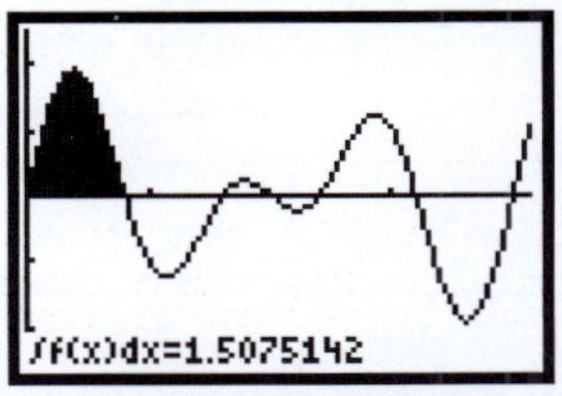

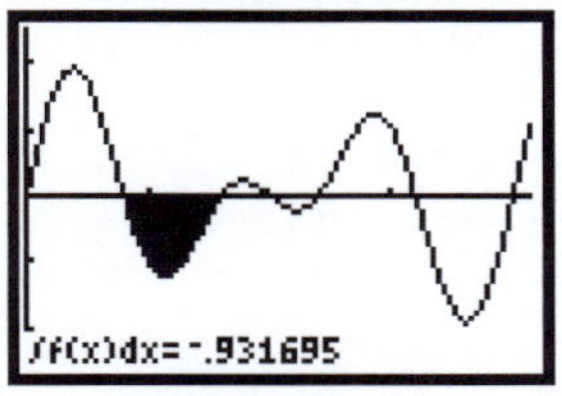

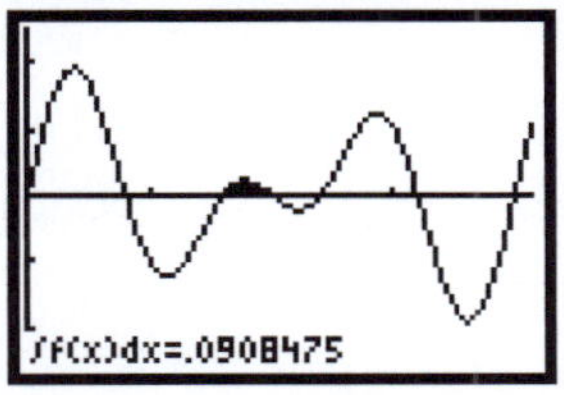


```
Ans→P
           1.507514162
abs(Ans)→Q
           .9316949906
Ans→R
           .0908474953
```

```
           .9316949906
Ans→R
           .0908474953
P+Q+R
           2.530056648
Ans*2
           5.060113296
```

(7) (b) The shortcut can be used to confirm any of the answers of Activity 5. Part (c) is shown here.

```
Plot1 Plot2 Plot3
\Y1=(X-1)³
\Y2=abs(Y1)
\Y3=
\Y4=
\Y5=
\Y6=
\Y7=
```

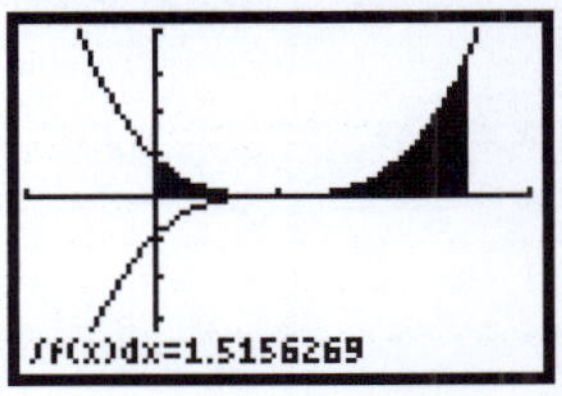

(8) (a)

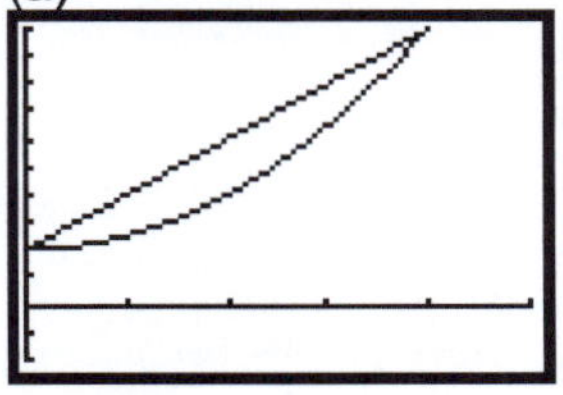

(b)

```
fnInt(Y2,X,1,3)
                    12
fnInt(Y1,X,1,3)
           8.333333333
12-Ans
           3.666666667
```

(c)

```
fnInt(Y2-Y1,X,1,
3)
           3.666666667
```

(d)

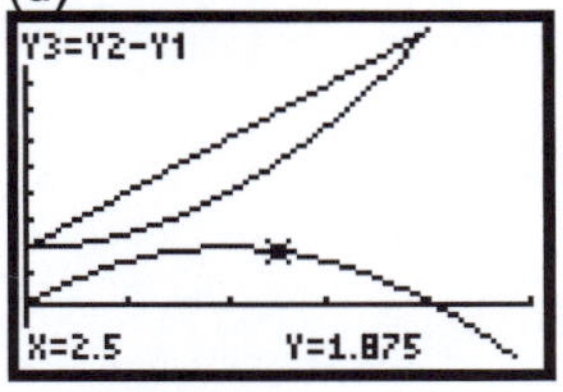

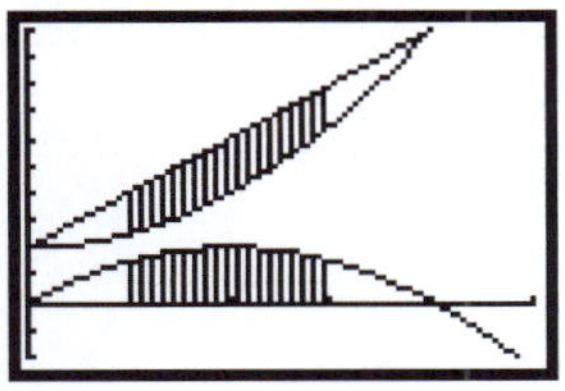

This shaded diagram provides visual confirmation that the area between Y2 and Y1 is the same as the area beneath Y2–Y1.

(9) (a)

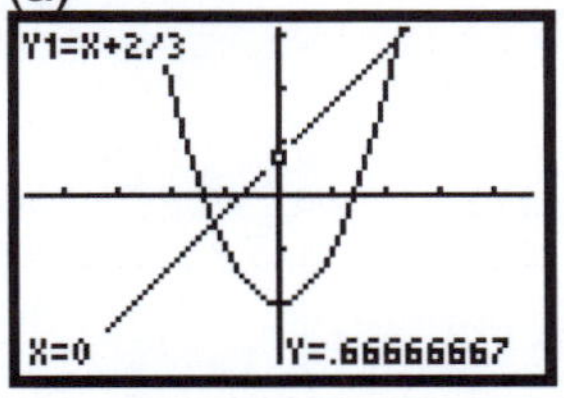

(b)

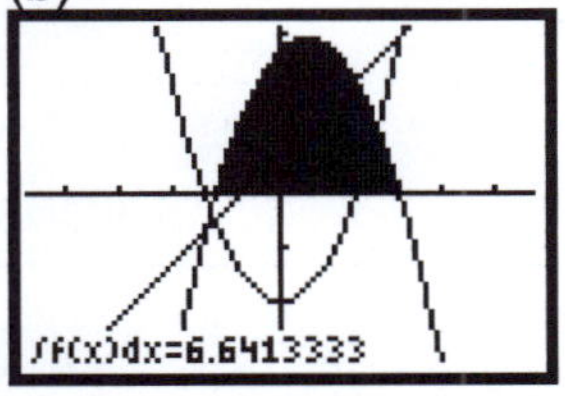

(c)

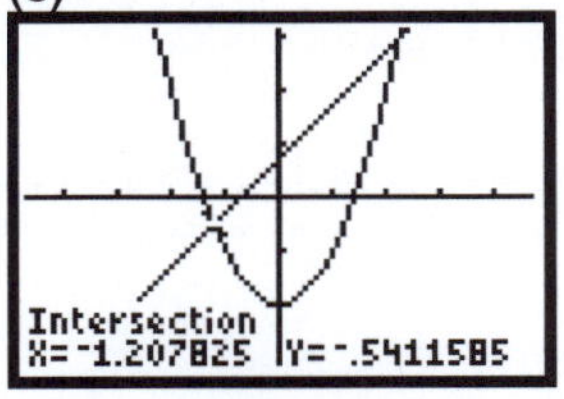

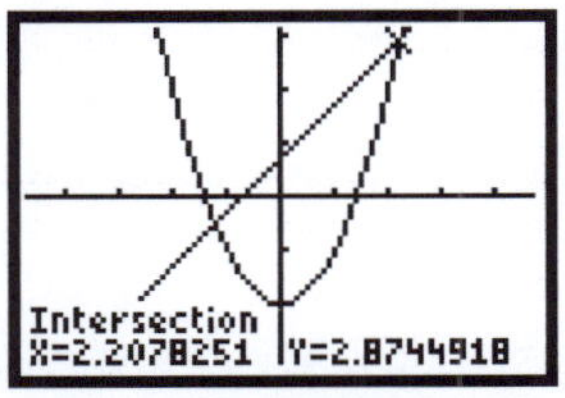


```
Ans→S
          -1.207825128
Ans→T
           2.207825128
fnInt(Y1-Y2,X,S,
T)
           6.641542163
```

(d) An alternative method for the final stage is shown here.

```
-S³/3+S²/2+8S/3-
(-T³/3+T²/2+8T/3
)
           6.641542163
```

(10) In each case it is best to start by drawing a graph.

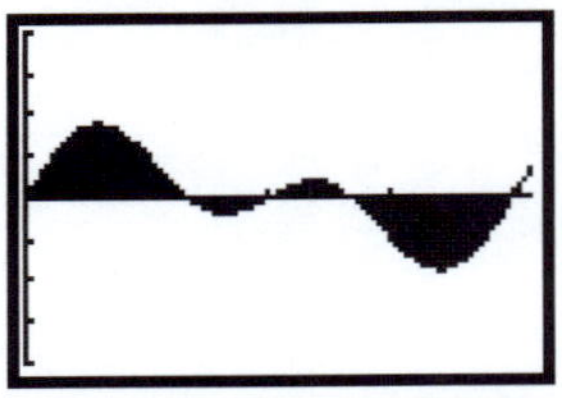

(a) There are four sections, the first two being symmetrical to the last two.

The method of Activity 7 is probably the easiest here.
Total area = 5.000 sq. units.

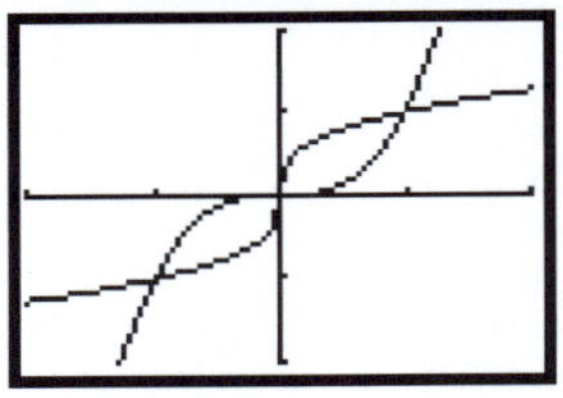

(b) There are two leaf shapes and the curves intersect at (1,1) and $(^-1,^-1)$.

The easiest method is probably to use **fnInt(Y2-Y1,X,0,1)** and to double the result.
Total area = 1.000 sq. unit.

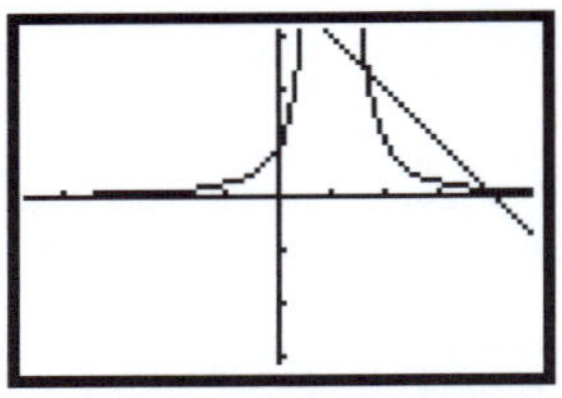

(c) You need to find where the curves intersect. (X=1.653 and X= 3.879).

If these values are stored in S and T, **fnInt(Y2-Y1,X,S,T)** can then be calculated.
The enclosed area is 1.563 sq. units.

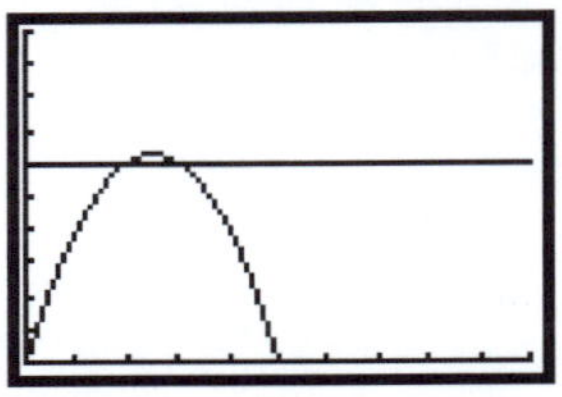

(d) The line and curve intersect at X=2 and X=3.

The method of Activity 9(a) works well here.

The area between the two lines is 0.167 sq. units.

So finally ...

$\int_a^b f(x)dx$ does usually equal the area between the curve $y=f(x)$, the x-axis and the lines $x=a$ and $x=b$.

However, there are exceptions, in particular where part or all of the area lies beneath the x-axis. If the area lies entirely beneath the axis, the integral equals the negative of the area. If the area lies entirely partly above and partly below the axis the integral produces a wrong answer.

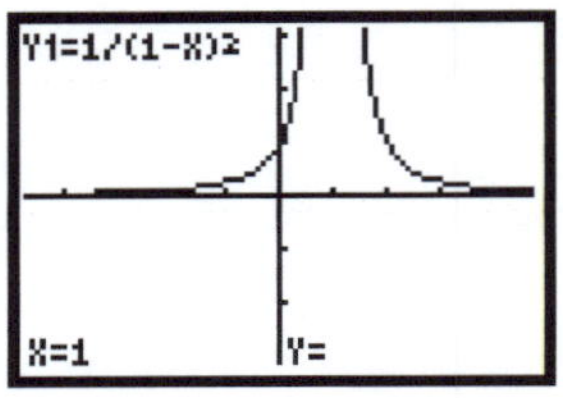

The function shown here exemplifies another type of exception. Here the function itself has no value at $x=1$, so an integral such as $\int_0^2 \frac{1}{(1-x)^2}dx$ cannot be evaluated. There is also no defined area between the curve the x-axis and the lines $x=0$ and $x=2$. Similar exceptions occur whenever the graph is not continuous in the domain being considered.

Unit 14 Volumes, means and changing rates

1) (a)

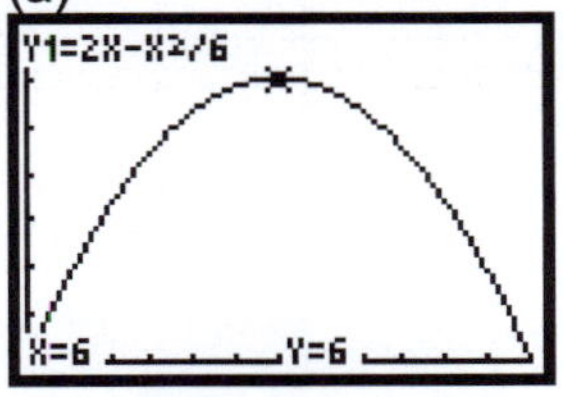

(b) The calculated mean, 3.67 (to 2 d.p.) lies within the range of daytime temperatures and seems a reasonable answer.

(c)

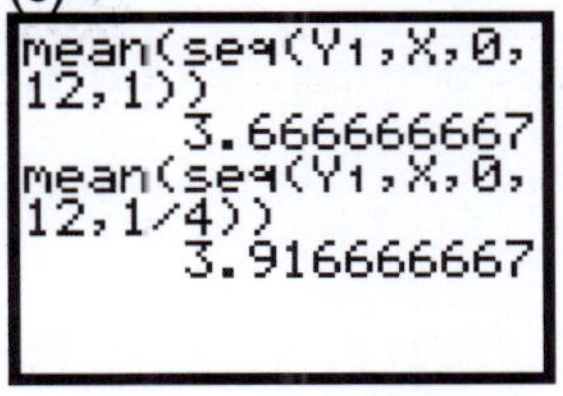

(d)

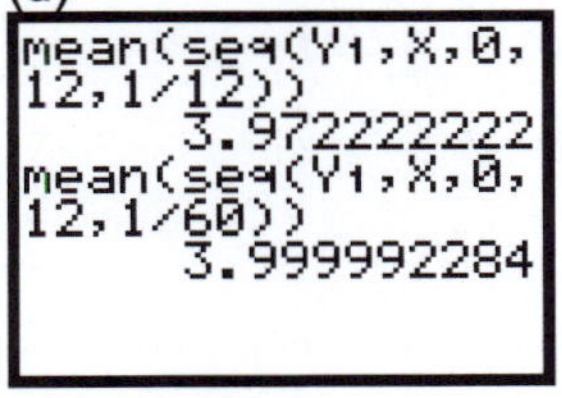

In the limit, as δx tends to zero, it seems that the mean temperature tends to 4 C.

(e)

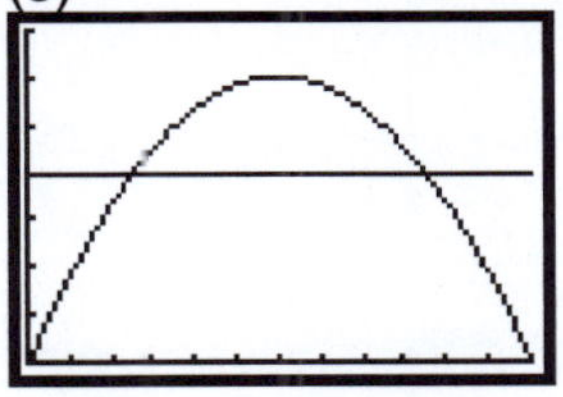

Brain stretcher

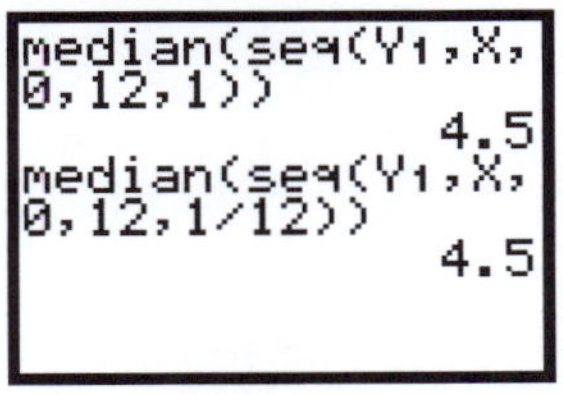

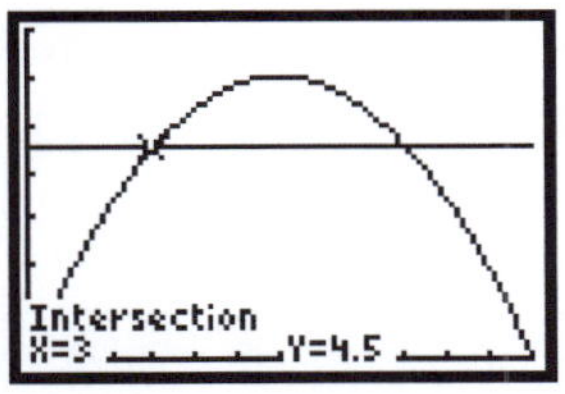

The median temperature is 4.5 whatever time interval is chosen. The median temperature occurs at x=3 and x=9, corresponding to times of 3 and 9 hours after sunrise. This result is predictable because the median is the middle value of a set of data i.e. there are as many values above it as below it.

As can be seen from the graph, in the period between x=0 and x=6 there are as many temperatures above the temperature at x=3 as there are below it. In the same way, for the period between x=6 and x=12 the median temperature occurs at x=9. So, if the whole period from x=0 to x=12 is considered, there must be as many temperatures above 4.5 as below it.

2)

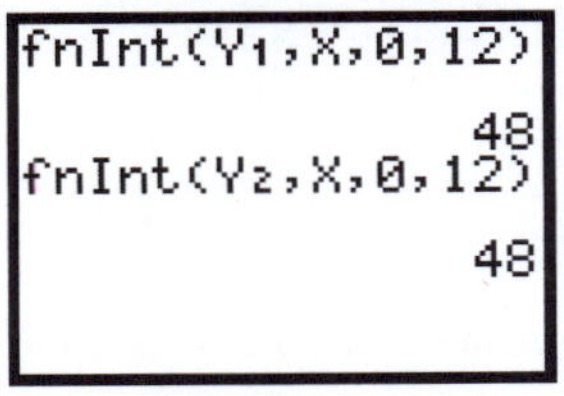

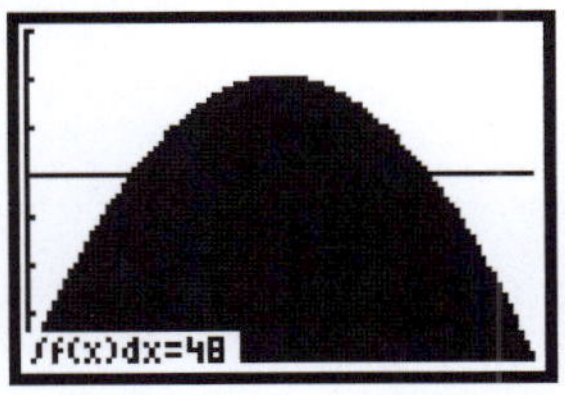

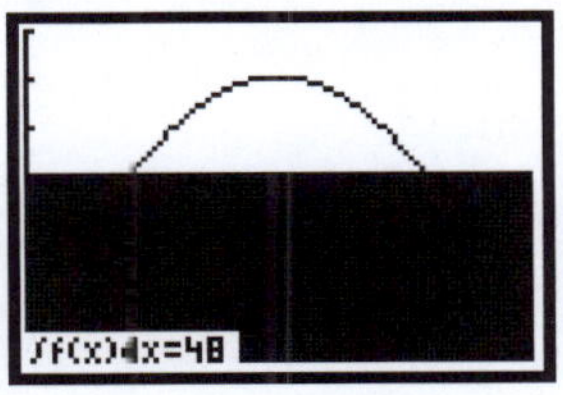

3) (a) The key presses required for program **MEANFN** are below.

PROGRAM: MEANFN	[PRGM] [◄] **1 MEANFN** [ENTER]
:Prompt A,B	[PRGM] [►] **2** [ALPHA] **A** [,] [ALPHA] **B** [ENTER]
:Disp fnInt(Y1,X,A,B)/(B–A)	[PRGM] [►] **3** [MATH] **9** [VARS] [►] **1 1** [,] [X,T,Θ,n] [,] [ALPHA] **A** [,] [ALPHA] **B** [)] [÷] [(] [ALPHA] **B** [–] [ALPHA] **A** [)]

Finally, press [2nd] [QUIT] to return to the home screen.

(b)

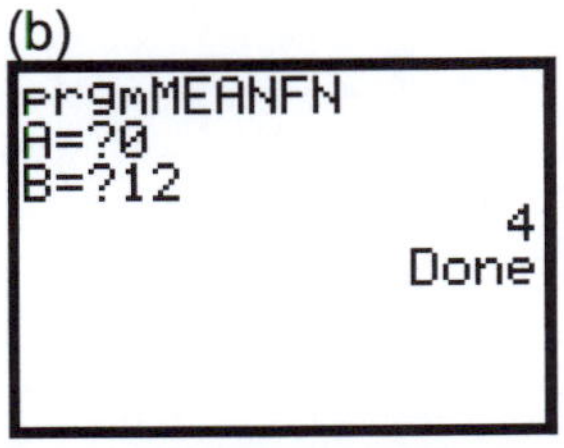

(c) The middle four hours of daylight are from x=4 to x=8 and the mean temperature is 5.8 (to 1 d.p).

4) (a) At t=2 there were 20 000 bacteria, but the mean number over the four-hour period was much more – just over 27 000.

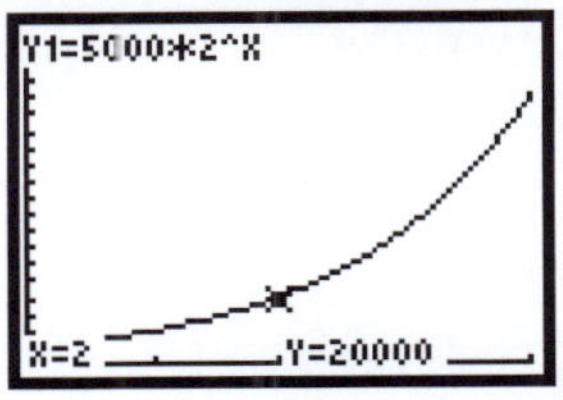

(b) You need to enter **Y1=sin(X)**.

```
prgmMEANFN
A=?0
B=?π/2
          .6366197724
                 Done
```

```
prgmMEANFN
A=?π/4
B=?3π/4
          .9003163162
                 Done
```

(5) (a)

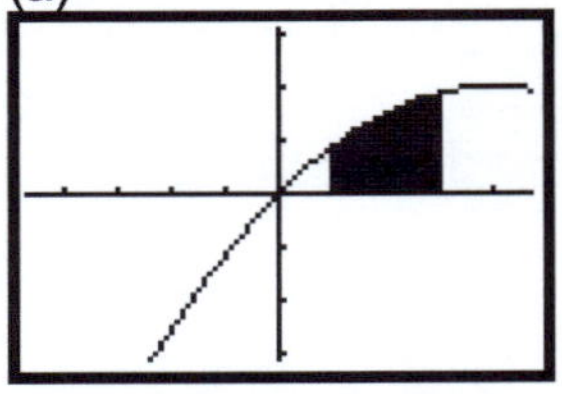

```
fnInt(Y1²,X,1,3)
*π
         13.89500251
```

(b) You can do this by entering **Y1=$\sqrt[3]{}$(X)** and carrying out the calculation shown here. The volume is just over 190 cubic units.

```
fnInt(Y1²,X,10,2
0)*π
         190.2775427
```

(6) (a)

```
fnInt(Y1²,X,-2,2
)*π
         33.51032164
4/3*π*2³
         33.51032164
```

(b) Any straight line through the origin will do. Here **Y1=X/2** has been used with X=4 giving values of *r*=2 and *h*=4. Both methods produce a volume of 5.3333π or $\frac{16}{3}\pi$.

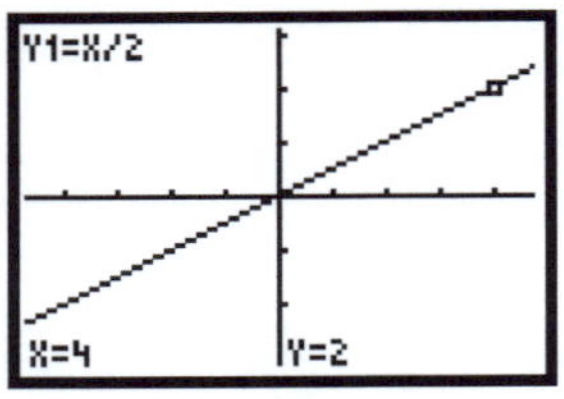


```
fnInt(Y1²,X,0,4)
         5.333333333
1/3*2²*4
         5.333333333
```

(c) You need to find the difference between the volumes swept out by Y1 and Y2. This can be done in two alternative ways as shown here.

```
(fnInt(Y1²,X,2,4
)-fnInt(Y2²,X,2,
4))*π
         39.47842118
fnInt(Y1²-Y2²,X,
2,4)*π
         39.47841887
```

(8)

```
fnInt(0.4T-T²/4,
T,0,1)
                 .12
fnInt(0.4T-T²/4,
T,1,2)
                 .02
```

```
fnInt(0.4T-T²/4,
T,2,3)
                -.58
```

The drug will cause the patient's temperature to rise slightly during the first two hours of treatment, before falling more rapidly during the third hour.

Notice that it is possible to use variables other than X within the **fnInt** command. For the displays shown here the calculator was displaying just 2 decimal places (using line 2 of the MODE menu).

(9) (a)

```
fnInt(3X²+2X+4,X
,0,.6)
                2.98
fnInt(3X²+2X+4,X
,.6,1.1)
                3.97
```

A and B are 2.98 m apart and B and C 3.97 m apart.

(b) Since $3X^2+2X+4$ represents the velocity, it is necessary to plot the antiderivative function to display a graph showing displacement. You can use Trace to check the positions of B and C.

Again for the displays shown here, the calculator was displaying just 2 decimal places.

```
Plot1 Plot2 Plot3
\Y1=3X²+2X+4
\Y2=fnInt(Y1,X,0
,X)
\Y3=
\Y4=
\Y5=
\Y6=
```

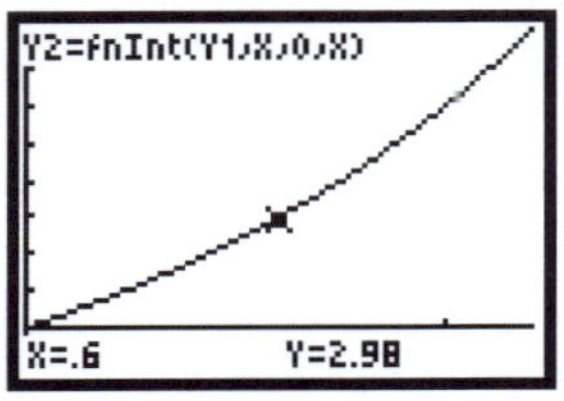

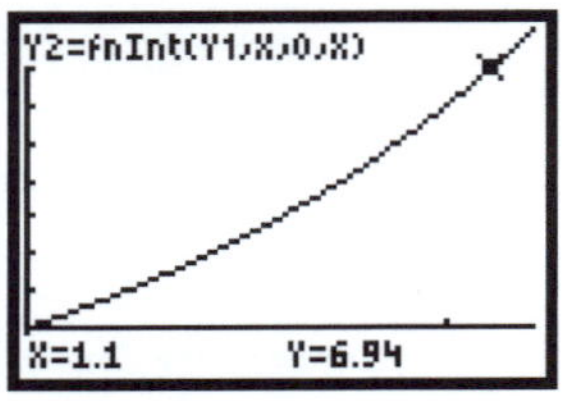

INDEX